Springer Texts in Business and Economics

Springer Texts in Business and Economics (STBE) delivers high-quality instructional content for undergraduates and graduates in all areas of Business/ Management Science and Economics. The series is comprised of self-contained books with a broad and comprehensive coverage that are suitable for class as well as for individual self-study. All texts are authored by established experts in their fields and offer a solid methodological background, often accompanied by problems and exercises.

Ramo Palalić • Claire Seaman •
Veland Ramadani • Léo-Paul Dana •
Gadaf Rexhepi

Sustainable Family Business

The Triple-Bottom-Line Approach

Springer

Ramo Palalić
Sultan Qaboos University
Muscat, Oman

Veland Ramadani
South East European University
Tetovo, North Macedonia

Gadaf Rexhepi
South East European University
Tetovo, North Macedonia

Claire Seaman
Queen Margaret Business School
Queen Margaret University
Musselburgh, UK

Léo-Paul Dana
ICD Business School Paris
IGS Group
Paris, France

ISSN 2192-4333 ISSN 2192-4341 (electronic)
Springer Texts in Business and Economics
ISBN 978-3-031-57998-1 ISBN 978-3-031-57999-8 (eBook)
https://doi.org/10.1007/978-3-031-57999-8

This Springer imprint is published by the registered company Springer Nature Switzerland AG
The registered company address is: Gewerbestrasse 11, 6330 Cham, Switzerland

To my parents, Ibrahim (1942–2022) and Zulejha (1950–2020), for their immense love and support

Ramo Palalić

To my family, especially David, with love and thanks for all the support

Claire Seaman

To my friends, Shqipe and Berat, who became my family—with boundless appreciation and respect

Veland Ramadani

To Michelle with thanks

Léo-Paul Dana

To my late uncle Nuri Rexhepi-Dana, in recognition of his contribution to my education

Gadaf Rexhepi

Preface

Family businesses' nature, legacy, long-term orientation, and strong community engagement are privileged considerations that distinguish them from non-family businesses. Such a combination of moral duties which is established on a healthy relationship with external stakeholders positions family businesses to dedicate lots of resources to societal well-being. This demonstrates that family businesses' existence goes beyond profit maximization. Yet, they maximize their contributions to socio-economic development through environmental and social channels of value delivery.

This preface introduces readership with the intricate relationship that family businesses have with sustainable practices, emphasizing the interdependence of social justice, environmental sustainability, and financial success. Therefore, family businesses have become important players in the changing global business arena, with a unique opportunity to promote sustainability and the Triple-Bottom-Line (TBL) concept.

This book content emphasizes family businesses' role in creating the long-term value through the sustainable business activities. Moreover, basic concepts in family businesses are presented stressing out the value chain creation through the sustainable competitive advantage. Family businesses' sustainable competitive advantage is gained through sustainable but complex orchestra of finance, human resource, marketing and public relations, research and development (R&D), supply chain management (SCM), global operations, and leadership.

Each of the book's theme is reflected in the integrated view of social, economic, and environmental factors, which is not simply a strategic imperative. Balanced and sustainable future is something that family businesses are leading in the mission of being stewards of legacy and community. Family businesses' community embeddedness and symbiotic life in the current posture of the global business scene make family businesses unavoidable factor in creating the future sustainability through triple-bottom-line approach.

Muscat, Oman — Ramo Palalić
Musselburgh, UK — Claire Seaman
Tetovo, North Macedonia — Veland Ramadani
Paris, France — Léo-Paul Dana
Tetovo, North Macedonia — Gadaf Rexhepi

"This book provides an in-depth and comprehensive analysis of sustainable practices in family businesses. Authored by well-recognized authors in the field, this work is a testament to the intricate relationship between family businesses and their sustainable development. The book is meticulously organized into nine insightful chapters, each focusing on a different aspect of sustainability within the family business context. From exploring the basic concepts and typologies of sustainable family businesses to delving into the specifics of sustainable finance, human resource management, marketing, and public relations, the authors leave no stone unturned. They also adeptly address operations and supply chain management, research and development, global operations, and leadership, all within the realm of sustainability. What sets this book apart is its focus on the triple-bottom-line approach, emphasizing the importance of social, economic, and environmental impacts in sustaining family businesses. The authors skillfully weave these three threads throughout the book, presenting a balanced view of how family businesses can thrive while contributing positively to society and the environment. This book is a must-read for scholars, practitioners, and students interested in the intersection of family business and sustainability."

—Andrea Caputo,
Editor of Strategic Change Journal
and Journal of Enterprising Communities
Professor at the University of Trento, Italy,
and University of Lincoln, United Kingdom

"This book combines the components of sustainability with family business. Each chapter identifies sustainable steps to maintain the family business and develop long-term success. This book is a must-

read for those in the area of family business and sustainability, which are 95% of the world's businesses."

—Dianne H.B. Welsh,
Hayes Distinguished Professor of Entrepreneurship;
Winner of 4 Lifetime Achievement Awards;
The University of North Carolina at Greensboro,
Greensboro, NC, USA

"This novel book merges the fundamental elements of sustainability with the intricacies of family business. A thorough exploration of each chapter uncovers practical and sustainable measures vital for the continuous growth and long-term prosperity of family firms. This inspiring masterpiece becomes an indispensable resource for professionals, researchers, and enthusiasts alike, providing invaluable insights and guidance to navigate the dynamic realm of family business while ensuring a sustainable future. This book is an absolute must-read for anyone involved in the captivating intersection of family business and sustainability."

—Wassim J. Aloulou,
Associate Professor of Entrepreneurship;
3rd Place Winner of the University
Excellence in Research Award;
Imam Mohammad Saud Islamic University,
Riyadh, Saudi Arabia

"This book offers an innovative and comprehensive perspective on managing sustainable family businesses. With a thoughtful and insightful approach, this work explores the complexities and challenges faced by family businesses, offering practical solutions and effective strategies to achieve sustainable success. This book is essential reading for entrepreneurs, business leaders, and academics seeking to understand and implement responsible business practices, aligning profit, people, and planet. A book that inspires and empowers, offering valuable insights to transform family businesses into engines of sustainable progress."

—Cristina Fernandes,
Assistant Professor at the University of Beira Interior,
Covilhã, Portugal and Loughborough University,
Loughborough, United Kingdom

"The book is an indispensable resource for anyone aspiring to build a family business that not only survives but flourishes across successive generations."

—Marina Dabić, PhD.,
Full Professor, Faculty of Economics & Business,
University of Zagreb
Trg J. F. Kennedyja 6, 10000 Zagreb, Croatia
Strategic Chair for Accreditation

Contents

About the Authors

Ramo Palalić is an Assistant Professor at the Management Department, College of Economics and Political Science, Sultan Qaboos University, Oman. Dr Palalić has authored and co-authored many articles in globally recognized journals like *Management Decision, International Journal of Entrepreneurial Behavior & Research, International Entrepreneurship and Management Journal*, and alike. Additionally, he has co-authored/co-edited several books and many book chapters in the field of entrepreneurship. Moreover, Dr. Palalić is serving as the *Co-Editor in Chief of Gestion 2000 Journal, Associate Editor of Asia Pacific Management Review (APMR), Enterprising Communities (JEC), and Heritage and Sustainable Development (HSD).* Additionally, he is an active reviewer and editorial board member in several well-established international journals.

Claire Seaman is a Professor Emerita in Enterprise and Family Business at Queen Margaret University, Edinburgh, UK. She has published over 30 journal articles and several books on family businesses in the UK, New Zealand, and Latin America and is a regular contributor to family business conferences worldwide. Prof. Seaman maintains close links with the business community and business organizations such as the Scottish Family Business Association, Family Business United, and local economic development agencies. She is a frequent speaker at gatherings of family businesses and professional advisors.

Veland Ramadani is a Professor of Entrepreneurship and Family Business at the Faculty of Business and Economics, South East European University, North Macedonia. His primary areas of research focus on entrepreneurship, family businesses, innovation, and sustainability. Dr. Ramadani has an extensive body of work to his name, having authored or co-authored over 140 research articles, 80 book chapters, 12 textbooks, and 28 edited books. His research contributions are notably featured in prestigious academic journals, including the *Journal of Business Research, Technological Forecasting and Social Change, Technology in Society, International Journal of Entrepreneurial Behavior & Research*, and *Review of Managerial Science,* among others. Furthermore, his research influence is underscored by his consistent inclusion among the World's Top 2% of the most influential researchers, as determined by Elsevier and Stanford University for the last four consecutive years. Dr. Ramadani has remained actively engaged in various projects, including those funded by the European Union and universities in Saudi Arabia, Kuwait, and the United Arab Emirates. Additionally, he regularly offers consultancy services to private companies.

Léo-Paul Dana is a Professor at ICD Business School Paris (France). He is also associated with the Chaire ETI at Sorbonne Business School. A graduate of McGill University and HEC-Montreal, he has served as Marie Curie Fellow at Princeton University and Visiting Professor at INSEAD. He has published extensively in a variety of journals.

Gadaf Rexhepi is Full Professor of Strategy and Innovation at South East European University, Republic of North Macedonia. He has published around sixty research articles in different peer and refereed journals, among which Business Strategy and the Environment, Journal of Business Research, Sustainable Development, Corporate Social Responsible and Environment Management, Review of Managerial Science, Industrial Marketing Management, etc. Most of his work has been in the field of CSR and Sustainability. He has served as a consultant for development of the Rector of South East European University. From 2020 he is a member of the Council of National Bank of the Republic of North Macedonia. He received the Award for Excellence 2016—Outstanding Paper by Emerald Group Publishing (Journal of Enterprising Communities: People and Places in the Global Economy).

1 Sustainable Family Business: Concept and Typology

Learning Objectives
After reading this chapter, you should be able to:

- Define family business from different perspectives.
- Understand the concept of sustainability.
- Understand triple-bottom-line dimensions.
- Know the differences between green family entrepreneurship and conventional family entrepreneurship.

Profile: Blake Mycoskie—Toms

Blake Mycoskie, the visionary behind TOMS, established the brand's One for One® model, aligning with TOMS' mission of leveraging commerce for societal betterment. This innovative model was crafted to advocate corporate responsibility and conscientious consumerism, ensuring that for each pair of shoes sold, TOMS would donate a new pair to an individual in need.

The genesis of this profit-driven enterprise with a philanthropic core stemmed from Blake's journey to Argentina in 2006. Witnessing the adversities faced by barefoot children—ranging from vulnerability to the inability to access education—inspired him. This initial concept of donating for each sale expanded beyond shoes to encompass crucial needs like eye care and safe water, facilitated through TOMS' eyewear and coffee sales.

Note: Selected sections of this chapter have been reused from the author's own work [Chapter: Nature of family business. In Ramadani, V., Memili, E., Palalic, R. and Chang. E. (2020). *Entrepreneurial Family Businesses* ©Springer, 2020].

R. Palalić et al., *Sustainable Family Business*, Springer Texts in Business and Economics, https://doi.org/10.1007/978-3-031-57999-8_1

Since its inception, TOMS has collaborated with numerous organizations, including UNICEF, Save the Children, Partners in Health, The Red Cross, Everytown, Faith in Action, March for Our Lives, among others. Blake's simple yet profound idea has evolved into a global movement, positively impacting over 100 million lives.

In 2018, Blake unveiled a new phase of TOMS' philanthropy by pledging a $5 million contribution to US-based organizations combatting gun violence—the most substantial corporate donation for this cause. This initiative included a nationwide campaign advocating Universal Background Checks, and distributing over 730,000 postcards to Congress. Within 15 days of this effort, the House passed H.R.8, mandating background checks for all firearm sales.

Blake's innovative business approach has garnered numerous accolades. From receiving the Secretary of State's Award of Corporate Excellence (ACE) in 2009 to joining The B Team led by Richard Branson in 2014, and earning Harvard's Next Generation Award in 2015, Blake has been recognized for his leadership and commitment to global health initiatives. His impact has extended to being honored with the Cannes Lion Heart Award in 2016, celebrating positive brand-driven influences.

In 2015, Blake established the Social Entrepreneurship Fund, supporting innovative, socially conscious companies across various industries, including technology, education, and food equality.

Before TOMS, Blake initiated five businesses, notably a prosperous campus laundry service. Amidst these ventures, he also participated in the CBS series, The Amazing Race, with his sister Paige, coming close to winning the grand prize.

An ardent reader, traveler, and golfer, Blake is fervent about motivating young individuals to contribute to a better future by integrating philanthropy into their pursuits. His vision revolves around fostering socially conscious businesses and consumers.

Source: Based on https://www.toms.com/us/corporate/blakes-bio.html

1.1 Family Businesses: Definitions and Specifics

1.1.1 Definition Approaches

Family businesses represent the most prevalent and ancient form of business organization, encompassing both small enterprises catering to local communities and large conglomerates operating across various industries and countries (IFC, 2008; Hisrich & Ramadani, 2017). Consequently, defining a family business proves to be a multifaceted challenge. The crux lies in the interplay between the family system and the business, as highlighted by Chua et al. (1999). When the founding editors of *Family Business*

Review asked, "What is a family business?" it became evident that while people generally grasp the concept, articulating a precise definition reveals the intricate nature of this phenomenon (Lansberg et al., 1988). Hoy and Verser (1994) observed that the editors intentionally refrained from providing a specific definition, opting instead to foster a dialogue within *Family Business Review* to delineate the field's boundaries. Manuscript submitters were encouraged to clarify the definition they were employing, enabling readers to make informed comparisons between studies.

Diaz-Moriana et al. (2019) have observed that there are five primary approaches to defining family businesses:

1. ***Circle models***. This approach involves circle models, which classify family businesses using the two and three-circle models. In the two-circle model, family businesses are seen as two closely interdependent systems, family and business, where the family system is based on emotions, inward-looking, long-term loyalty, conservative stance to change, and ensuring the equilibrium of the family remains intact; while the business system is oriented toward task accomplishment, results and performance. Three-circle model (Tagiuri & Davis, 1996), by using three spheres of influence in the family business—owners, managers (employees), and family members, introduces the bivalent attributes of the family businesses.
2. ***Defining family business by behavior.*** This approach is based on Chua et al. (1999) who require defining family business based on distinctive behavior. They contend that the family business is defined by its specific behavior and "not on the basis of the components of family involvement, but by how these components are used to pursue the family's vision" (Chua et al., 1999, p. 27).
3. ***Defining family business by the degree of family involvement***. Based on this approach, definitions of family businesses accentuate the family participation in the business and their significant influence and control over the business's future strategic direction, and the owner's intentions for active involvement of multiple family generations in the business operations.
4. ***Familiness.*** This approach applies the resource-based view (RBV) to the family business context. Chrisman et al. (2003) defined the familiness construct as "resources and capabilities related to family involvement and interactions" (p. 468). The authors argue that the family business system creates resources and by influencing them, they become valuable, rare, and highly inimitable by competitors.
5. ***Family Influence on Power, Experience, and Culture (F-PEC) scale and Substantial Family Influence (SFI).*** According to F-PEC scale, the family business definition should allow for heterogeneity, and instead of evaluating particular traits or behaviors, F-PEC measures the family based on its influence on the company, such as family ownership, voting control, family management, family employment, generational transfer, and interdependent subsystems, and is multiple inclusive (Astrachan et al., 2002). SFI evaluates the full power of the family by measuring the family's share in equity (zero to one); the family share of the supervisory board (zero to one); and the share of the top management team (zero to one) (Klein Astrachan & Smyrnios, 2005).

When examining the various methods of defining family businesses, Memili and Dibrell (2019), drawing from an analysis of 82 distinct empirical studies, identified the following criteria commonly used in family business definitions:

- Ownership
- Management
- Control
- Generational aspects
- Subsystems
- Perception

It is widely acknowledged that there is no universally accepted or single definition of a family business, "which is exclusively applied to every conceivable area, such as to public and policy discussions, to legal regulations, as an eligibility criterion for support services, and to the provision of statistical data and academic research" (Mandl, 2008, p. 1). Astrachan et al. (2002) have pointed out that "a definition of family is often missing…this notable absence poses problems, particularly in an international context where families and cultures differ not only across geographical boundaries but also over time" (p. 167). The broad concept of a family business encompasses any enterprise where a majority of ownership or control is held by a family, and where two or more family members are directly involved (Brockhaus, 2004). A family business constitutes a doubly intricate system, integrating both the aspects of business and family. These systems intersect, constituting dynamic organisms that evolve and undergo changes, each possessing its distinct history, challenges, strengths, weaknesses, opportunities, and threats. Family members engaged in the business operate within the framework of business tasks and the family system simultaneously. Consequently, conflicts may arise due to the distinct rules, roles, and requirements inherent in each system. While families may have their unique styles of communication and conflict resolution, which may be suitable within the family context, it does not necessarily translate to effective resolution of business disputes. Entry into the family system is typically through birth, adoption, or marriage, with membership assumed to be permanent. In contrast, entry into the business system is contingent on experience and opportunities. Conflicts can emerge when issues from one system are transferred to the other (Bowman-Upton, 2009; Gashi & Ramadani, 2013).

Defining a family business necessitates an exploration of its uniqueness, posing the fundamental question of "What is unique?" It extends beyond the mere fact of family members owning or managing a business. The distinctive quality of a family business lies in how the model of ownership, governance, and succession management significantly influences its objectives, strategies, structure, and the formulation, design, and implementation of business activities (Chua et al., 1999; Mandl, 2008; Xi et al., 2015).

Poza and Daugherty (2013) outline specific characteristics that qualify a business as a family business, including (a) ownership control of 15% or more by two or more family members; (b) strategic influence by family members in the management, whether through active involvement, cultural creation, advisory roles, board membership, or as active shareholders; and (c) a concern for family relationships, encompassing the aspiration for continuity across generations. Expanding on these features, Poza and Daugherty introduce additional aspects: (a) the presence of the family; (b) the convergence of family, management, and ownership, with its zero-sum (win-lose) tendencies, making family businesses particularly susceptible during succession, especially in the absence of firm growth; (c) unique sources of competitive advantages, such as a long-term investment horizon, arising from the interplay of family, management, and ownership, particularly when family unity is high; and (d) the owner's aspiration to perpetuate the business within the family, emphasizing the goal of business continuity across generations.

As per the European Commission's criteria (2009), a company, regardless of its size, is classified as a family business if:

- The majority of decision-making rights lie with the natural person(s) who founded the company, those who have acquired its share capital, or their spouses, parents, children, or children's direct heirs.
- The majority of decision-making rights, whether direct or indirect, involve at least one representative of the family or a relative who is formally engaged in the company's governance.
- For listed companies, the definition of a family business applies if the individual who established or acquired the company (share capital), or their families or descendants, holds 25% or more of the decision-making rights determined by their share capital.

According to Chua et al. (1999), a family business is "a business governed and/or managed with the intention to shape and pursue the vision of the business held by a dominant coalition controlled by members of the same family or a small number of families in a manner that is potentially sustainable across generations of the family or families" (p. 25).

Family businesses differ from non-family ones in many ways. According to Mandl (2008), in the center of the firm in family businesses is family, which formally or informally, directly or indirectly influences the firm; their main objectives are both economic and non-economic, respectively, sustainability/long-term family income (stability) as well as family satisfaction; their business orientation is the satisfaction of internal and external stakeholders (mainly family, clients, employees, local community); the style of management is value-driven, emotional and goal alignment, they compete on quality, reputation, long-term relationships, etc. Alternatively, in the center of non-family businesses are the owner(s) or managers; their main objectives are only economic (quick profits/growth); their business orientation is the satisfaction of owners/shareholders; the style of management is facts-and-figures-driven, rational and use agency control mechanisms, etc.

In broad terms, a family business can be described as a business entity owned and governed by members of the same family. This type of business often involves the employment of family members and operates with the expectation that younger generations within the family will eventually assume control and leadership responsibilities, following the precedent set by their elder family members (Ramadani et al., 2017).

1.1.2 Economic Contribution of Family Businesses

Family businesses, as the prevailing form of business organization, make a substantial contribution to the economies of various countries. On a global scale, family businesses makeup approximately 80–98% of all businesses, generate about 75% of the gross domestic product (GDP) in most countries, and employ more than 75% of the workforce worldwide. In the United States, they account for 49% of the GDP, employ 80% of the workforce, and are responsible for generating 86% of new jobs (Poza, 2010; Poza & Daugherty, 2013). In the United Kingdom, as of 2017, there were more than 4.8 million family businesses, constituting 85% of all private sector enterprises. These family businesses employed 13.4 million individuals, which represents 50% of private sector employment and 38% of all employment while contributing £598 billion to the GDP (IFB Research Foundation, 2019). In Canada, approximately 80% of companies listed on the Toronto Stock Exchange are categorized as family businesses (Gulzar & Wang, 2010). Data from the Spanish Family Business Institute reveal that family businesses form 85% of the Spanish business sector, contribute to 70% of the national GDP, and employ 70% of the private sector workforce (Fernández-Olmos et al. in 2016). Similarly, in Saudi Arabia, family businesses make up 95% of all companies and are responsible for around 30% of the GDP (Rahman et al., 2017). These statistics underscore the significant role played by family businesses in community development, job creation, and the enhancement of people's quality of life.

1.1.3 Advantages and Disadvantages of Family Business

Drawing from the literature, several advantages and disadvantages of family businesses have been identified (Dana & Ramadani, 2015; Ramadani & Hoy, 2015). The advantages of family businesses include:

- *Continuity and reinvestment:* Family members serve as both owners and managers of the business, and ownership can potentially be passed down to future generations. Consequently, most of these businesses tend to reinvest their profits back into the company, ensuring its sustainability and long-term success.
- *Inherent commitment:* The employment of family members means that the business engages individuals who have a vested interest in the success of the enterprise. When challenges arise, family members are often more deeply committed and concerned about the company's well-being compared to non-family employees.
- *Societal benefit:* Family businesses provide advantages not only to the family itself but also to society at large. In addition to employing family members, these

businesses create job opportunities for other individuals with the skills and qualities needed to contribute to the business.

- *Customer relations:* Family businesses often foster close and enduring relationships with their customers. This familiarity and trust can lead to the establishment of long-term customer connections. Customers perceive the family name associated with the company as a symbol of trust, as they believe that the family is unlikely to compromise its reputation through subpar, unethical, or illegal practices. This trust can significantly contribute to the business's stability and longevity.

Besides advantages, family businesses have their disadvantages as well. Some of these drawbacks are highlighted below:

- *Family-related issues:* Family businesses can become a source of numerous problems within the family itself, leading to issues such as gambling, anxiety, worries, or even substance abuse, such as drug and alcohol addiction. It is quite uncommon for family emotions not to intersect with business affairs at some point, potentially causing internal conflicts and challenges.
- *Nepotism and misuse of positions:* Family business managers may encounter difficulties in refraining from employing their relatives, even when these family members lack the necessary qualifications or skills required for the business. In many instances, these family members have been known to abuse their positions within the business simply because they belong to the family.
- *Succession challenges:* Family members, particularly parents who have held leadership positions within the business for many years, might find it difficult to accept the need for a transition of power. They may resist stepping aside to allow the next generation or other family members to take over the business, potentially missing out on new ideas and innovations that these successors could bring to the family business.

1.2 The Concept of Sustainable Business

1.2.1 Definition and Triple-Bottom-Line Dimensions

Sustainable development is perhaps one of the most important issues of our time. The development of the concept of sustainability started in 1987 when the Brundtland Commission put forward a "global agenda for change" in order to "advance the shared understanding and shared spirit of responsibility so clearly needed in a divided world" (Hisrich et al., 2020). According to this report, sustainable development includes the development "that meets the needs of the present without compromising the ability of future generations to meet their needs" (Brundtland, 1987).

It is assumed that in order to achieve sustainable development, family business can play an important role. Family business can help us be both sustainable and developmental (Ramadani et al., 2022). In general, *sustainable business* focuses on preserving nature, supporting life and community (sustainability), and pursuing perceived opportunities to bring new products, processes, and services with profit goals (entrepreneurial action), where profit will be used to bring economic and non-economic benefits to individuals, the economy and society (Adapted from Hisrich et al., 2020).

Usually, the business action is driven by knowledge and motivation. Those with greater knowledge of the natural environment—the physical world, including land, biodiversity, and ecosystems—are more likely to notice changes in that environment that create a host of opportunities than those with less knowledge. Nonetheless, it is crucial not to underestimate the significance of (business/entrepreneurial) understanding concerning markets, technologies, and/or the exploitation of opportunities. Devoid of this knowledge, the realization of opportunities for sustainable development is improbable (Shepherd & Patzelt, 2011).

A business action to be considered as "sustainable," it must bring benefits for the family business owners, other people, and society/environment. Family business owners not only have the capacity to generate economic wealth for themselves but can also exert a substantial impact on broader development. Beyond personal financial gains, they have the potential to create benefits for others, encompassing economic, environmental, and social dimensions. In the literature, these multifaceted benefits are often referred to as the ***triple-bottom line***, signifying the interconnected considerations of economic prosperity, environmental sustainability, and social well-being. This concept underscores the idea that the impact of family businesses extends beyond financial success, contributing to a more holistic and sustainable approach to overall development (Fig. 1.1).

Economic benefits for others can be, for example, creating employment opportunities, providing better quality products and services, increasing the standard in the countries where family businesses operate, as well as increasing revenues in the state budget. Murphy Oil Corporation is a good example of the sustainable economic impact of businesses. Murphy Oil Corporation is a Fortune 500 company that addresses problems in El Dorado, Arkansas, such as declining industry, declining population, and declining school enrollment, in 2007 announced that it would donate $50 million for a scholarship program for local students, through the El Dorado Promise program. The initiative featured a component offering scholarships to students over the next two decades. One year following the program's introduction, there was a noteworthy 18% surge in university enrollments. Subsequently, after 2 years, the community experienced a 4% rise in school enrollments. The local college observed an even more substantial 16% increase in enrollment during the same period. Additionally, the community witnessed an influx of families from over 28 states and ten foreign countries who relocated to El Dorado. As a result of the El Dorado Promise program, an estimated 2000 students have enrolled in college, highlighting the positive impact of the initiative on higher education participation (Saylor Academy, 2012; Wann, 2020).

Environmental benefits are a result of activities such as reducing air pollution, improving air quality, improving the quality of drinking water, and improving other living conditions. Swedish giant IKEA has invested in sustainability throughout its

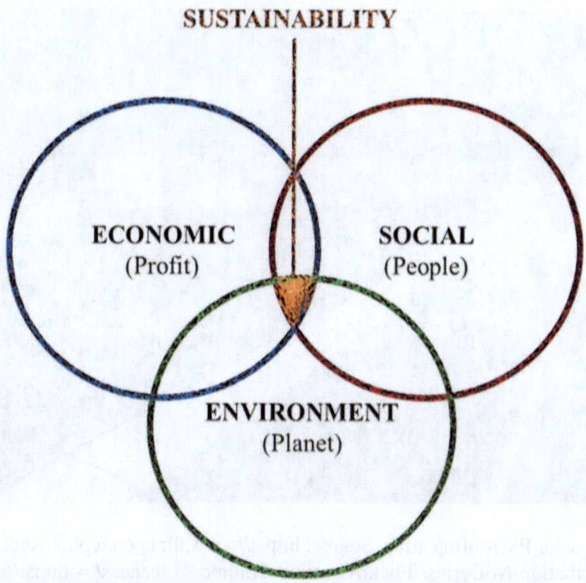

Fig. 1.1 Triple-bottom-line dimensions. *Source:* Authors

business operations, including the things customers can easily see and the things they cannot. It starts with its supply chain, where the Swedish furniture maker ensures approximately 50% of its wood from "sustainable foresters" and 100% of its cotton from farms that meet Better Cotton standards, which include reductions in water and energy use, as well as in the use of chemical fertilizers and pesticides. Since 2009, IKEA has invested more than $2.6 billion in renewable energy, owning and operating 416 wind turbines and installing approximately 750,000 solar panels on IKEA buildings (Fig. 1.2). In 2017, IKEA generated renewable energy equivalent to 73% of the energy they used in their operations. IKEA is working to become a circular business, with which it wants all its products to be designed from the beginning to be reused, repaired, resold, and recycled (IKEA, 2022).

Social benefits include improving child survival rates, life expectancy rates, improving higher education, creating equal opportunities, creating better working conditions, stopping abuse of children and women in workplaces, etc. (Ramadani et al., 2022). TOMS Shoes serves as an exemplary illustration of a company dedicated to generating social benefits for others. Established in 2006 by Blake Mycoskie, TOMS Shoes is driven by a singular mission: to enhance the lives of children by supplying them with much-needed footwear. The manufacturing of these shoes takes place in Argentina and China, adhering to fair labor practices and striving for minimal environmental impact. Continuous monitoring of factories is conducted by TOMS and independent third-party auditors to ensure ethical production standards. TOMS shoes are made available for purchase both online and in retail stores globally. A distinctive feature of TOMS' business model is its commitment that, for every pair of shoes sold, the company will donate a second pair to a

Fig. 1.2 IKEA solar PV rooftop array. Source: https://www.flickr.com/photos/teia/26236016284, licensed as Attribution-NoDerivs/ Flickr/Creative Common Licences/ – https://www.flickr.com/photos/teia/26236016284

child in need, extending its impact to regions such as Argentina, South Africa, and various other countries worldwide. This buy-one-give-one approach reflects TOMS Shoes' dedication to creating positive social change through its business operations. They invite consumers around the world to experience first-hand the sense of social contribution of TOMS Shoes, under the "buy one, give one" model. Since 2006, over 50 million pairs of shoes have been donated to children in need (Upneja, 2016).

Nevzat Aydin—A Classic and Rare Example of a Sustainable Family Business Owner

Nevzat Aydin, the owner of "Yemeksepeti," an online platform for food product sales, has made the admirable decision to share a portion of the proceeds, a substantial $27 million, from the sale of his business with his 114 employees. The acquisition of "Yemeksepeti" by the German company "Delivery Hero" amounted to $589 million. Mr. Aydin recognizes the invaluable contributions of his employees to the company's success and believes they deserve this recognition. Each employee has received a monetary reward based on their tenure with the company, ensuring that no one received less than $215,000.

Mr. Aydin emphasizes that achievements are not the result of overnight endeavors but are attained through persistent hard work and unwavering dedication. He acknowledges that while he may have led the business, it is the collective efforts of his employees that have been instrumental in its success, emphasizing their integral role in the company's journey.

Source: http://money.cnn.com/2015/07/28/news/companies/turkish-ceo-staff-bonus-millions/

The concept of sustainable business may seem overly optimistic at first glance, resembling an idealistic perspective. However, this viewpoint aligns with the notion of considering business actions as a tool, akin to a hammer that can be wielded for positive purposes, such as constructing a community center, or for negative purposes, acting as a weapon to cause harm. In a study involving 83 entrepreneurs, researchers discovered that while the majority of business owners held positive attitudes toward the natural environment, there were instances where these values were set aside to pursue opportunities that inadvertently caused harm to the environment (Shepherd et al., 2013). Despite these challenges, there is a growing belief that many individuals today are motivated to leverage business actions in support of the natural environment and communities, aiming to create and develop benefits for others.

Sustainable Development Goals

Economic goals

Goal 8. Promote sustained, inclusive, and sustainable economic growth, full and productive employment, and decent work for all.
Goal 9. Build resilient infrastructure, promote inclusive and sustainable industrialization, and foster innovation.
Goal 10. Reduce inequality within and among countries.
Goal 12. Ensure sustainable consumption and production patterns.

Societal goals

Goal 1. End poverty in all its forms everywhere.
Goal 2. End hunger, achieve food security and improved nutrition, and promote sustainable agriculture.
Goal 3. Ensure healthy lives and promote well-being for all at all ages.
Goal 4. Ensure inclusive and equitable quality education and promote lifelong learning opportunities for all.
Goal 5. Achieve gender equality and empower all women and girls.
Goal 7. Ensure access to affordable, reliable, sustainable, and modern energy for all.
Goal 11. Make cities and human settlements inclusive, safe, resilient, and sustainable.
Goal 16. Promote peaceful and inclusive societies for sustainable development, provide access to justice for all and build effective, accountable, and inclusive institutions at all levels.

Environmental goals

Goal 6. Ensure availability and sustainable management of water and sanitation for all.
Goal 13. Take urgent action to combat climate change and its impacts.

Goal 14. Conserve and sustainably use the oceans, seas, and marine resources for sustainable development.

Goal 15. Protect, restore, and promote sustainable use of terrestrial ecosystems, sustainably manage forests, combat desertification, and halt and reverse land degradation and halt biodiversity loss.

Overall goal

Goal 17. Strengthen the means of implementation and revitalize the Global Partnership for Sustainable Development.

Source: United Nations (https://sdgs.un.org/goals).

1.2.2 Green Family Entrepreneurship: Definition and Characteristics

The definition of green family entrepreneurship is a challenging task for most researchers. The concept is relatively new, dating back to the 1990s of the last century. Recently, the interest in green entrepreneurship has been growing. In the literature, different terms are used for green entrepreneurship, such as eco-entrepreneurship, environmental entrepreneurship, ecological entrepreneurship, and the like, making it even more difficult to clearly define this concept (Schaper, 2016). Further, many scholars increase the complexity of the definitions by including the ethical, social, or environmental motivations, thus coming very close to the definition of sustainable entrepreneurship. It is worth underlining that the term "sustainable" strongly includes three aspects, social, economic, and environmental, while the term "green" mainly focuses on the environmental aspect (Zubeltzu-Jaka et al., 2018). Researchers have raised several questions, the most common of which are these two: (1) What activities can be included in the "green" part of entrepreneurship? and (2) What are the main characteristics of green entrepreneurs? As can be seen, the literature has not yet provided clear answers to these questions (OECD, 2011).

In the following, we will present some definitions of the green entrepreneur that would enable us to create an overview of this new concept of entrepreneurship. According to Saari and Joensuu-Salo (2019), Green entrepreneurship is a distinct category within the entrepreneurial landscape, with a primary objective of developing and implementing solutions to address environmental issues while simultaneously driving societal change to prevent harm to the environment. Further, some even argue that green entrepreneurship represents an entirely new business paradigm, rather than merely a subgroup of "traditional" entrepreneurship. This perspective arises from the fact that green entrepreneurs operate with more expansive motivations beyond solely introducing eco-friendly products and services tailored for a niche market.

Green family entrepreneurship includes family business activities that consciously deal with an environmental problem or need through the realization of entrepreneurial ideas with a high-risk level, which has a net positive effect on the natural environment and at the same time is financially sustainable (Koester, 2001). Green family entrepreneurship refers to an economic activity where the products, services, production methods, or organizational practices have a positive impact on the environment (Gevrenova, 2015). This positive impact can manifest in various ways, including:

- Offering products or services that, when consumed, prompt a shift in consumer behavior, thereby diminishing negative impacts on the environment.
- Striking a balance between the ecological and economic objectives of the company.
- Introducing innovative ecological solutions to address issues linked to the production and consumption of products and services.
- Identifying business models that, when implemented, contribute to sustainable economic development.
- Recognizing new market opportunities aligned with societal demands and evolving lifestyles.

Furthermore, green family entrepreneurship includes those family businesses whose core business model addresses environmental issues—improves energy or resource efficiency, reduces emissions of harmful gases, reduces waste or pollution, protects or restores ecosystems, promotes local culture, and supports the community (Miller et al., 2017). Wood (2021) listed several green business ideas, such as:

- Energy audits and green consultancy: Working with businesses or individuals to implement strategies for reducing their carbon emissions.
- Environmental law: Providing legal services to protect the environment or hold entities accountable for violating environmental laws.
- Eco restaurants: Opening a restaurant that uses only locally sourced food and sustainable products and processes.
- Development of wind energy: Building wind farms to produce sustainable energy.
- Recycled furniture and household goods: Creating furniture and household goods using recycled materials.
- Organic spa: Opening a spa that uses organic products, avoiding harsh chemicals harmful to both skin and the environment.
- Green architecture: Designing new homes using sustainable methods.
- Green venture capitalism: Investing in helping other people get their green business ideas off the ground.

According to Amankwah and Sesen (2021), a green family entrepreneur can be considered a person who is committed to converting his traditional business to green by adopting environmentally friendly production technologies or someone who enters a business to produce green products and services, to supply green raw materials and create green jobs, whether in traditional manufacturing or construction sectors or in new sectors such as renewable energy and energy efficiency.

Green Family Entrepreneurs Typology

Several types of green entrepreneurs can be found in the literature. The main types of green family business entrepreneurs are ad hoc green entrepreneurs, innovative opportunists, ethical mavericks, and visionary champions (Fig. 1.3).

Ad hoc ***green entrepreneurs*** (accidental) are driven purely by financial gain and whose contribution to sustainability is completely unintentional and not as a request from the authorities. An illustrative example of such an entrepreneur is Tom Archer in the enduring UK BBC Radio 4 drama, The Archers. As the son of dedicated and prosperous organic farmers, Tom decides to establish his own venture specializing in the production of organic pork. The supportive sociocultural background provided by his parents, who are deeply committed to organic farming, lays the foundation for Tom's entrepreneurial journey. Driven by a desire for independence, Tom aspires to demonstrate his capability as a successful businessman in his own right. This fictional character embodies the essence of green family entrepreneurship, leveraging both familial expertise and a commitment to sustainable practices in the pursuit of business success. Organic pig production happens to be a complementary activity to the rest of the family business. It is not a product whose development has been specifically encouraged by government subsidies or driven by environmental regulation or consumer demand.

Innovative opportunists are driven by financial motives, and after noticing a green opportunity in existing markets, decide to exploit it (e.g., recycling of old refrigerators, hybrid cars, and bio-derivatives). This type of entrepreneur is mainly

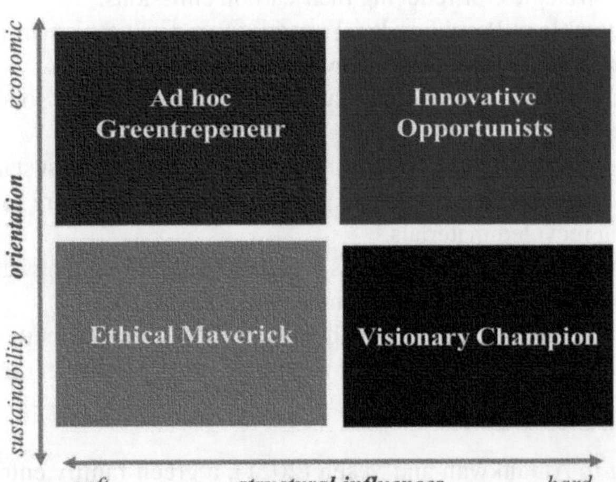

Fig. 1.3 Typology of green entrepreneurs. Source: Based on Taylor and Walley (2004)

influenced by strong structural drivers, such as legal regulation. A possible example of this type of entrepreneur is Evan Connell, of the Connell Group, who built a factory to safely recycle old refrigerators. In accordance with European Union environmental legislation, it is mandated that all chlorofluorocarbons (CFCs) must be recovered. In response to this directive, Connell implemented a series of new technologies, developed in collaboration with Manchester Metropolitan University. These technologies facilitated the removal of harmful CFC gases from the foam layer inside old refrigerators, ensuring their safe disposal as scrap. This initiative aligns with environmental regulations and reflects a commitment to sustainable practices in handling and recycling materials containing ozone-depleting substances.

Ethical mavericks are those entrepreneurs whose orientation to sustainability is motivated by the values of their friends, networks, and past experiences, rather than by the motive to change the world. These entrepreneurs tend to create small businesses in special markets (niche markets) that rarely grow (and when they do, they grow at a small rate) due to the lack of financial interest on the part of the founder (e.g., health food stores, as an alternative business). Examples of this type of green entrepreneur are the founders of the vegetarian cafe and health food store "On the 8th Day" in Manchester, England.

Visionary champions are those entrepreneurs who embrace a transformative, sustainable orientation, and are motivated by a combination of their values and financial return (e.g., pure natural hair and skin products). A classic example of this type is Anita Roddick of The Body Shop.

Green Family Entrepreneurship and Conventional Family Entrepreneurship (Start-Up)

Green family entrepreneurship and conventional family entrepreneurship (start-up), in addition to the differences related to the search for opportunities, creativity and innovation, rational management of resources, and the tendency to face risks, also have their own differences. The authors Saari and Joensuu-Salo (2019) have identified some relevant differences between these two concepts. According to them, the primary distinction between conventional family entrepreneurship and green family entrepreneurship lies in the underlying logic of value creation. Conventional family entrepreneurship typically focuses on contributing to economic growth, fostering regional development, and potentially enhancing local economies. In contrast, green family entrepreneurship pertains to businesses with a core objective of reducing their environmental footprint and minimizing the impact of their operations on the natural environment. Essentially, green family entrepreneurship emphasizes the ecological dimension of sustainability as a crucial component of their value creation. In conventional family entrepreneurship, the primary, and often sole, driver is the generation of economic value. Conversely, in green family entrepreneurship, the economic aspect is viewed as a means to achieve other values that are closely associated with the environment and society. While conventional family entrepreneurs are increasingly responding to sustainability demands stemming from environmental regulations and the growing consumer interest in greener products and services, green family entrepreneurs set themselves apart by not only prioritizing financial performance but also by integrating the environmental dimension into their core business strategy. They actively seek out and leverage business opportunities to mitigate or eliminate harm to the natural environment, thus demonstrating a commitment beyond profit and a dedication to broader societal and environmental values.

Green Family Entrepreneurship Barriers

Green family entrepreneurship development encounters several barriers, including financial struggles, regulatory barriers, market limitations, technological constraints, supply chain issues, competition, growth challenges, investor hesitance, and a need for better support systems (Makki, 2020; Thanasi-Boçe et al., 2023). Overcoming these barriers requires collaborative efforts involving governments, investors, businesses, and communities to create an environment conducive to green entrepreneurship. Table 1.1 summarizes the main barriers to green family entrepreneurship development.

Table 1.1 Main barriers to green family entrepreneurship development

Main barriers to the green family entrepreneurship development
• *Costs associated with green initiatives*: Implementing environmentally friendly practices may involve significant upfront costs, deterring businesses from adopting green measures. • *Lack of knowledge and subjective awareness in the market*: Limited awareness and understanding of green initiatives among consumers and businesses can hinder their adoption. • *Lack of investors and involvement of the private sector*: Insufficient financial support and engagement from private investors may impede the growth and scalability of green initiatives. • *The short-term mindset of investors*: Investors focusing on short-term returns may be hesitant to invest in sustainable projects that often yield long-term benefits. • *Government regulations*: Stringent or unclear regulations can pose challenges for businesses seeking to adopt and implement green practices. • *Cultural differences*: Differing cultural attitudes toward sustainability may impact the acceptance and integration of green initiatives in various regions. • *Dominant industries*: Industries with established practices and a reluctance to change may resist adopting green initiatives. • *Lack of incentive and support mechanism*: The absence of sufficient incentives and support structures may discourage businesses from embracing sustainable practices. • *Bureaucracy*: Cumbersome bureaucratic processes and red tape can slow down the implementation of green initiatives. • *Lack of technology*: Inadequate access to or development of green technologies may limit the feasibility of sustainable practices. • *Lack of research and development (R&D)*: Insufficient investment in research and development can hinder the creation of innovative green solutions. • *Lack of cooperation between interested parties*: Collaboration between various stakeholders, such as businesses, government bodies, and communities, is essential for successful green initiatives.

Source: Based on Makki (2020)

Case Study Apple, Chinese Suppliers, and Environmental Problems

In 2011, Apple faced allegations from a Chinese environmental group, the Institute of Public and Environmental Affairs in Beijing, accusing both Apple and its Chinese suppliers of causing environmental contamination and posing threats to public health. The group's comprehensive 46-page report documented numerous instances of environmental damage linked to Apple and its suppliers across the globe.

The report, based on on-site visits to multiple factory locations, highlighted shortcomings in the proper disposal of hazardous waste by Apple's supplier factories. It revealed that 27 suppliers associated with Apple were found to have environmental problems. Earlier in the same year, Apple acknowledged that 137 workers at a Chinese factory near Suzhou had sustained serious injuries due to exposure to a toxic chemical used in the production of iPhone screen glasses.

In response to the accusations, Apple's spokesperson emphasized the company's commitment to maintaining high standards of social responsibility throughout its supply chain. Apple claimed to conduct regular audits and closely monitor the factories within its supply chain. The spokesperson stated, "Apple is committed to driving the highest standards of social responsibility throughout our supply chain. We require that our suppliers provide safe working conditions, treat workers with dignity and respect, and use environmentally responsible manufacturing processes wherever Apple products are made."

Supply chain experts noted that larger, brand-name companies like Apple generally exhibit better monitoring and auditing practices within their supply chains compared to smaller companies in China. Despite improvements in some areas, experts acknowledged that labor violations and the improper discharge of toxic waste remained significant challenges in many locations.

Questions:

1. Should the environmental damage caused by Apple's suppliers be of concern to Apple's management? What are the risks and benefits of having suppliers in China?
2. Should Apple try to work with the Institute of Public and Environmental Affairs, other US companies with suppliers in China, the Chinese government, or the US government to address the issue of environmental damage caused by suppliers in China? Give concrete suggestions!

Source: Based on Barboza (2011).

Questions for Discussion

- Define family business based on different approaches!
- What are the advantages and disadvantages of family businesses?
- Define sustainable business!

- Explain the triple-bottom line!
- What is green family entrepreneurship?
- What are the main differences between green and conventional family entrepreneurship?
- Explain the typology of green family entrepreneurs!

Keywords
- Ad hoc green entrepreneurs
- Economic benefits
- Environmental benefits
- Ethical mavericks
- Family business
- Green family entrepreneurship
- Innovative opportunists
- Non-family business
- Social benefits
- Sustainable business
- Triple-bottom line
- Visionary champions

References

Amankwah, J., & Sesen, H. (2021). On the relation between green entrepreneurship intention and behavior. *Sustainability, 13*(13), 7474. https://doi.org/10.3390/su13137474

Astrachan, J. H., Klein, S. B., & Smyrnios, K. X. (2002). The F-PEC scale of family influence: A proposal for solving the family business definition problem. *Family Business Review, 15*(1), 45–58.

Barboza, D. (2011). *Apple suppliers causing environmental problems, Chinese Group Says*. New York Times.

Bowman-Upton, N. B. (2009). *Transferring management in the family-owned business*. U.S. Small Business Administration.

Brockhaus, R. H. (2004). Family business succession: Suggestions for future research. *Family Business Review, 17*(2), 165–177.

Brundtland, G. H. (1987). *Our common future*. Oxford University Press.

Chrisman, J. J., Chua, J. H., & Steier, L. (2003). Editorial: An introduction to theories of family business. *Journal of Business Venturing, 18*(4), 441–448.

Chua, H. J., Chrisman, J. J., & Sharma, P. (1999). Defining the family business by behaviour, *entrepreneurship. Theory and Practice, 23*(4), 19–39.

Dana, L.-P., & Ramadani, V. (2015). *Family businesses in transition economies*. Springer.

Diaz-Moriana, V., Hogan, T., Clinton, E., & Brophy, M. (2019). Defining family business: A closer look at definitional heterogeneity. In E. Memili & C. Dibrell (Eds.), *The Palgrave handbook of heterogeneity among family firms* (pp. 333–374). Palgrave Macmillan.

European Commission. (2009). Family business. Retrieved February 04, 2020, from https://ec.europa.eu/growth/smes/promoting-entrepreneurship/we-work-for/family-business_en

Fernández-Olmos, M., Gargallo-Castel, A., & Giner-Bagües, E. (2016). Internationalization and performance in Spanish family SMES: The W-curve. *BRQ Business Research Quarterly, 19*(2), 122–136.

Gashi, G., & Ramadani, V. (2013). Family businesses in Republic of Kosovo: Some general issues. In V. Ramadani & R. Schneider (Eds.), *Entrepreneurship in the Balkans: Diversity, support and prospects*. Springer.

Gevrenova, T. (2015). Nature and characteristics of green entrepreneurship. *Trakia Journal of Sciences, 13*(2), 321–323.

Gulzar, M., & Wang, Z. (2010). Corporate governance and non-listed family-owned businesses: Evidence from Pakistan. *International Journal of Innovation, Management and Technology, 1*(2), 124–129.

Hisrich, D. R., Peters, M., & Shepherd, D. (2020). *Entrepreneurship* (11th ed.). McGraw Hill.

Hisrich, D. R., & Ramadani, V. (2017). *Effective Entrepreneurial Management*. Springer.

Hoy, F., & Verser, T. G. (1994). Emerging business, emerging field: Entrepreneurship and the family firm. *Entrepreneurship: Theory and Practice, 19*(1), 9–23.

IFB Research Foundation. (2019). *The state of the nation: The UK family business sector 2018–19*. Oxford Economics.

IFC. (2008). *Family business governance handbook*. International Finance Corporation.

IKEA. (2022). *Designing for a circular future*. Available at: https://www.ikea.com/global/en/our-business/peopleplanet/designing-for-a-circular-future/. Accessed May 11, 2024

Klein, S. B., Astrachan, J. H., & Smyrnios, K. X. (2005). The F-PEC scale of family influence: Construction, validation, and further implication for theory. *Entrepreneurship Theory and Practice, 29*(3), 321–339.

Koester, E. (2001). Green entrepreneur handbook: The guide to building and growing a green and clean business.

Lansberg, I. S., Perrow, E. L., & Rogolsky, S. (1988). Family business as an emerging field. *Family Business Review, 1*(1), 1–8.

Makki, A. A. (2020). Barriers to green entrepreneurship: An ISM-based investigation. *Journal of Risk and Financial Management, 13*(11), 249.

Mandl, I. (2008). *Overview of family business relevant issues*. Austrian Institute for SME Research.

Memili, E., & Dibrell, C. (2019). *The Palgrave handbook of heterogeneity among family firms*. Palgrave Macmillan.

Miller, D. T., Dannals, J. E., & Zlatev, J. J. (2017). Behavioral processes in long-lag intervention studies. *Perspectives on Psychological Science, 12*(3), 454–467.

OECD. (2011). *Entrepreneurship at a glance*. OECD Publishing.

Poza, E. J. (2010). *Family business* (3rd ed.). Cengage Learning.

Poza, E. J., & Daugherty, M. S. (2013). *Family business* (4th ed.). Cengage Learning.

Rahman, H., Galván, R., & Martínez, B. (2017). Impact of family business on economic development: A study of Spain's family-owned supermarkets. *Journal of Business and Management Sciences, 5*(4), 129–138.

Ramadani, V., Agarwal, S., Caputo, A., Agrawal, V. & Dixit, J.K. (2022). Sustainable competencies of social entrepreneurship for sustainable development: Exploratory analysis from a developing economy. *Business Strategy and the Environment*, 31(7), 3437–3453.

Ramadani, V., Bexheti, A., Rexhepi, G., Ratten, V., & Ibraimi, S. (2017). Succession issues in Albanian family businesses: An exploratory research. *Journal of Balkan and Near Eastern Studies, 19*(3), 294–312. forthcoming.

Ramadani, V., & Hoy, F. (2015). Context and uniqueness of family businesses. In L.-P. Dana & V. Ramadani (Eds.), *Family businesses in transition economies* (pp. 9–37). Springer.

Saari, U. A., & Joensuu-Salo, S. (2019). Green entrepreneurship. In F. W. Leal, A. Azul, L. Brandli, P. Özuyar, & T. Wall (Eds.), *Responsible consumption and production: Encyclopedia of the UN sustainable development goals*. Springer.

Saylor Academy. (2012). *An introduction to sustainable business*. E-Book.

Schaper, M. (2016). Understanding the green entrepreneur. In *Making Ecopreneurs* (pp. 27–40). Routledge.

Shepherd, D., & Patzelt, H. (2011). Sustainable entrepreneurship: Entrepreneurial action linking 'what is to be sustained' with 'what is to BeDeveloped'. *Entrepreneurship: Theory and Practice, 35*(1), 137–163.

Shepherd, D., Patzelt, H., & Baron, R. (2013). 'I care about nature, but…': Disengaging values in assessing opportunities that cause harm. *Academy of Management Journal, 56*(5), 1251–1273.

Tagiuri, R., & Davis, J. (1996). Bivalent attributes of the family firm. *Family Business Review, 9*(2), 199–208.

Taylor, D. W., & Walley, E. E. (2004). The green entrepreneur: Opportunist, maverick or visionary? *International Journal of Entrepreneurship and Small Business, 1*(1/2), 56–69.

Thanasi-Boçe, M., Kurtishi-Kastrati, S., Ramadani, V., & Zuferi, R. (2023). Sustainable entrepreneurship in North Macedonia: Challenges and perspectives. In V. Ramadani, S. Kjosev, & B. S. Sergi (Eds.), *Entrepreneurship development in the Balkans: Perspective from diverse contexts* (pp. 197–211). Emerald Publishing Limited.

Upneja, A. (2016). TOMS: Impact and effectiveness of "buy one, give one" model. Retrieved January 10, 2022, from https://yaleglobalhealthreview.com/2016/07/08/toms-impact-and-effectiveness-of-buy-one-give-one-model/

Wann, T. (2020). *Murphy Oil to fund "El Dorado Promise" despite closing Ark. headquarters*. Available at: https://www.knoe.com/content/news/Murphy-Oil-continues-to-fund-El-Dorado-Promise-as-headquarters-in-El-Dorado-close-570291141.html. Accessed May 11, 2024

Wood, M. (2021). 39 green business ideas for sustainable entrepreneurs. Retrieved January 10, 2022, from https://www.nerdwallet.com/article/small-business/sustainable-business-ideas

Xi, J. M., Kraus, S., Filser, M., & Kellermanns, F. W. (2015). Mapping the field of family business research: Past trends and future directions. *International Entrepreneurship and Management Journal, 11*(1), 113–132.

Zubeltzu-Jaka, E., Erauskin-Tolosa, A., & Heras-Saizarbitoria, I. (2018). Shedding light on the determinants of eco-innovation: A meta-analytic study. *Business Strategy and Environment, 27*(7), 1093–1103.

2 Sustainability and Competitive Advantages in Family Businesses

Learning Objectives
After reading this chapter, you should be able to:

- Define competitive advantage.
- Define sustainable competitive advantage.
- Address the importance of sustainable competitive advantage.
- Understand tools and mechanisms of sustainable competitive advantage.

Profile: Bosch GmbH

Well-known global brand, Robert Bosch GmbH corporation, the third ranked family business in Germany, was founded in 1886 by Robert Bosch as "Workshop for Precision Mechanics and Electrical Engineering" in Stuttgart. It is one of the market leaders nowadays in advanced manufacturing and mobility industry. It employs over 400,000 employees worldwide. It has its headquarters in Stuttgart, and subsidies across the Europe, North and South America, Asia, and Australia.

From the very beginning, the business was characterized by the innovation and the social commitment (to resolve societal and peoples' problems). The modern corporation its history started as a repair shop for electrical equipment, but very soon the workshop had its two inventions. The first one happened when he was asked to build a "magneto ignito" for a stationary machine, in which he (Robert Bosch) improved the existing one. This was the milestone to make a massive production of magneto ignition during that time, and Bosch became the only supplier of that parts for the automotive

[1]Written by: Balqees Al Zadjali (Sultan Qaboos University).

R. Palalić et al., *Sustainable Family Business*, Springer Texts in Business and Economics, https://doi.org/10.1007/978-3-031-57999-8_2

industry. In 1902, a better solution was presented by Bosch's head engineer, Gottlob Honold, in the form of a high-voltage magneto ignition system with spark plug. This product helped Bosch establish itself as a top global supplier to the automotive industry. He was the "chief innovator with the right instinct."

Bosch soon had its manufacturing premises across Germany and Europe, especially in the UK. From that time, Bosch was existing in the UK for more than 120 years, and more than 50 decades in Turkey.

The company continued to have an innovative project even up to now. In 1990s, with the beginning of globalization, Bosch became a very known brand, which established its name on innovative products and right solutions. Its competitive advantage is based on a corporate innovation strategy that leads to the long-term sustainability.

One of the great examples where Bosch showed sustainable innovation, has happened in 2001, in the north American market. Bosch has proved its strategic innovation and market segmentation in North America, which seemed to be very challenging. It is a famous example of the Project called: "Bosch CS20 Circular Saw."

In 2001, Bosch confronted a mature and seemingly saturated market, but it succeeded in introducing a product that kept up with its high-quality image. Bosch outperformed competitors like DeWalt in North America, gained entry into big-box retailers, and offered targeted profit margins at a competitive price. Through the application of Strategyn's Outcome-Driven Innovation (ODI) methodology, Bosch was able to pinpoint a segment of circular saw primarily finish and advanced carpenters that accounted for over 30% of the user group which had 14 unmet needs out of 85 possible desired outcomes.

Bosch created the CS20 circular saw by incorporating cutting-edge technologies including the Direct Connect system, which attaches an extension cable directly to the saw, decreasing the risk of severing the cord while in use and enhancing user visibility.

When the CS20 was first released in 2004, it immediately jumped to the top of sales and user reviews. Popular Science even rated it one of the year's most inventive items. By focusing on specific client needs, Bosch's strategy not only made it easier for the firm to enter the market but also maintained a strong presence, allowing it to compete with industry leaders and diversify its product offerings.

The outcome-based strategy did a lot to do Bosch' sustainable competitive advantage in this market. Bosch's Vice President of Tools and Services, Andrew Reed, said: "*Focusing on customer outcomes enabled us to uncover over a dozen opportunities for improvement in the very mature circular saw market, many of which were previously unknown to our development team.*"

He also made a statement: *"The product that resulted from this insight has generated more interest and more initial sales than any other product we have created to date."* In addition, Randall Coe, Director of Product Development, said: *"It is highly unlikely that we would have focused on these outcomes, defined this feature set, or made the right trade-off decisions without having the benefit of the outcome-driven analysis."*

The bottom line is that the right innovation and the right corporate strategy lead to sustainable competitive advantage. Your say?!

Bibliography

https://familybusinessindex.com/ Accessed on 01-01-2024
https://www.bosch.com/company/our-history/ Accessed on 01-01-2024
https://strategyn.com/jobs-to-be-done/bosch-case-study/#form Accessed on 01-01-2024

2.1 Introduction

Family business is a very specific venture that differs from non-family business in many ways. One difference, for example, lies in firms' *emotional history* that non-family businesses do not have or do not have in that way as family business does. Hence, the development and growth of family firms is as hard as the dissolution of family businesses. Non-family firms can be sold out without any or little emotions attached, however, a family enterprise is flooded with emotional binds in which owners, as well as family members, are emotionally fastened to the enterprises' business (Ramadani et al., 2020). Taking into consideration future perspectives of family businesses (FBs), some FBs are in the latest trends that will help them in terms of profit, but also it will help the society and the Planet.

Family firms like, every small and medium enterprises (SMEs), endeavor to survive, develop, grow, and eventually sustain their growth in the market. To be the leader in the market, firms need to produce or provide goods and services that will be more appealing to customers for which those customers will be willingly pay higher prices than for the products of the competitors' products. This is the case of getting better than others, and this moment in business is called competitive advantage. Competitive advantage is relevant regardless of the firm's size or industry, whether the firm is local, regional, international, multinational, or global. In this chapter, we will define nature of competitive advantage and sustainable competitive advantage. Moreover, societal, economic, and environmental impacts of sustainable competitive advantage will be discussed. The chapter will be closed by a brief discussion on a summary of competitive advantage followed by the implication of TBL approach to family firms. Lastly, a case study is provided at the end of the chapter that will enhance learning and discussion.

2.2 Nature of Competitive Advantage

Competitive advantage happens when a firm has a better financial performance over its competitors in the same industry. Another perspective of competitive advantage is if a company's performance is above the industry average (Ahmad et al., 2022; Rothaermel, 2015). When a firm outperforms its competitors for prolonged period of time, then the firm attains a **sustainable competitive advantage**.

Not all firms have this opportunity to be the best over a prolonged time due to strategic decisions made internally, and of course, external factors challenge the competitive advantage of the firms. Thus, in this context, we say that competitive advantage of the firm is "*relative*" (Rothaermel, 2015, p. 5). Why relative? Because is not easy to get advantage over competitors, which means your firm's financial performance must be better than theirs. The financial performance is better because your sales have increased over a period of time. The sales are increased due to quality products, which can be reflected in any of products' qualities (performance, design, physical features, etc.). The increase in sales and achieve competitive advantage can be obtained in two different ways (but hard to implement them).

First, to win a competitive advantage over a firm's competitors the firm must have (design) or develop such products or services that are superior in terms of value for targeted customers. In this case, the price of superior products and services will be higher, which ultimately after the products are sold the sales will be higher. Second, products or services will be sold in big volumes if the firm has such products or services that are similar to its competitors, but sold at lower prices. In the first case, we can observe that the firm developed differentiated products while in the second case products or services are mostly standardized and similar to competitors, but costs are contained so that they can be sold at lower prices. Here, firms must have kind of trade-offs to get the right strategic positioning in the market. To be a low-cost firm, or upscale firm competing in the market, whether it is an SME or a family firm, the rules of the market game are the same. Looking at the example of Apple in early 2000, to survive in the market it had to attain competitive advantage in the prolonged period. By innovative approach, Steve Jobs created differentiated products and informed customers about them so that sales increased tremendously. However, Steve Jobs did not stop at first or second year of Apple's success or 2 years after his company achieved a competitive advantage.

Rather he immensely searched for new innovations at Apple and urged people to follow his path of success. In the the case of Apple, the trade-off was to be upscale firm which will provide high quality and innovative products while containing costs. Once the trade-off is set, it will then be easier for managers to use their available resources to gain a competitive advantage. For instance, if there is a decision that a family firm has decided to go for the **cost leadership strategy** (focusing on low-cost products), then no need for a high investment to create differentiated or unique products. Rather, drafting the strategy that will be reflected in products similar in quality and value to competitors' one, is the key focus. However, if the family firm decides to go for a **differentiation strategy** (focusing on development and delivering differentiated-unique features of products) (Rothaermel, 2015), then managers also know what to do, how, and what resources will be mobilized.

Additionally, a very important segment in drafting strategy (in attaining competitive advantage) nowadays is (IT) technology (Dehning & Stratopoulos, 2003; Lumkin et al., 2002). It is inevitable and paramount tool that will help incumbent family firms to gain and sustain competitive advantage. The technology seems too expensive at the very beginning of the investment, but it can be very crucial in cost leadership strategy for the long term (if the family firm decides on this type of strategy).

2.3 Societal Impact of Sustainable Competitive Advantage

Think of a company that is committed to the multifaceted promotion of social development and growth. What approaches would make their contribution truly meaningful? Well, significant social impacts do not happen overnight and cannot be expected to produce immediate high-level results. Well, it does not happen overnight, and thus it cannot be expected to have a high-immediate societal impact.

The societal concept was brought by Kotler and his associates (Kotler et al., 2020a, p. 13) throughout their books on marketing. Kotler and his team emphasized the balance between the firm's profit, consumers' wants, and the welfare of the society. The company wants profit, which will be the source of a continuous improvement of existing products and services, and to develop new ones, while firms' people will be motivated to perform necessary jobs. Similarly, consumers quest for valued products that will be at least satisfactory to them. And society wants to be treated fairly so that the firm provides positive outcomes to the whole society.

It is a kind of circle, which contributes to some mutual relationships in which the firms should play a fair game. If any of the stages of the game is played unfairly, the competitive advantage may be jeopardized. Products provided by the firms will be bought by consumers, which brings profit to the firm. Firms should take care of needs of the society in a way that will be beneficial to the firm and society on a fair basis.

Society does not directly affect the firm's profit, but society is very influential on consumers, and it can give a positive or negative signal to consumers about the firm's play. If positive, consumers remain loyal, and new consumers are being attracted. If negative, loyalty might be decreased with no attraction of new ones.

The bottom line is if a firm truly contributes to society, there will be a positive effect in attaining competitive advantage over a long term, and might be sustainable if other internal strengths are kept unchanged (firm's core competencies, know-how, innovation, etc.). nonetheless, the firm should be ambidextrous in this interplay.

2.4 Economic Impact of Sustainable Competitive Advantage

The previous interplay in this triangular circle (Fig. 2.1) could be a very important step toward competitive advantage. Assuming that all goes well within this interplay, the firm can earn trust in the society and be more motivated to contribute beyond the local level. It will create products and services that could be exported to some other regions which at the state level increases the overall country's economic outlook.

Coming back to the local society, the firm should intend to create products or services that will resolve some problems from different perspectives of consumers. On the other hand, the family firm can bring the economic value to consumers (society) in a sense that will provide products and services which will be economically more affordable than the ones imported from other countries. However, the family firm can internationalize and help other consumers internationally by providing them with similar products and services that bring an economic value to that society.

The current business context for family firms across the globe, in terms of technology and digitalization, is on their side. Savvy family firms should use this opportunity to explore, conquer, and flood other markets with their innovative products (implementing differentiated strategy) or similar products and services to competitors (implementing low-cost strategy).

Digital age is a great opportunity to obtain and sustain competitive advantage for the long term if accompanied by the smart use of IT possibilities. We have witnessed and still see that firms, family or non-family firms, which had some good IT

Fig. 2.1 Savvy firm attaining Sustainable advantage through IT and digitalization opportunities. Source: Authors' compilation

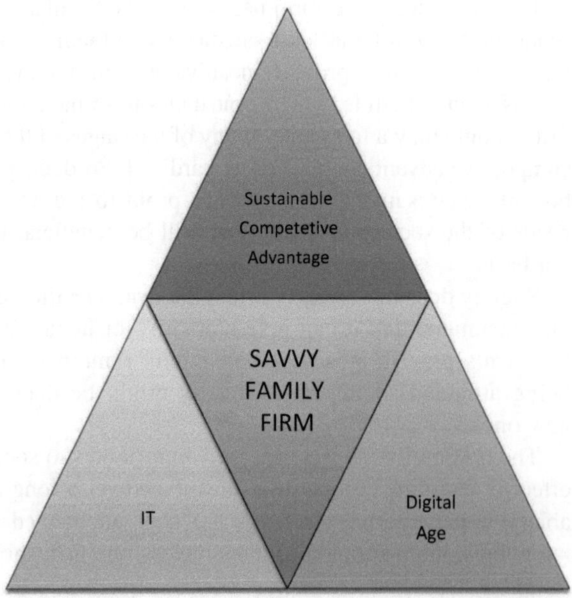

platforms they achieved competitive advantage very quickly in this pandemic. If there is a problem (COVID-19), there is an opportunity too. Hence, the family firms should be IT advanced because it is being well paid, especially in turbulent times (Knudsen et al., 2021).

2.5 Environment and Economic Sustainable Competitive Advantage

Family firms (FFs) usually possess very unique qualities in comparison to non-filmy businesses. These qualities help FFs to gain a strong sustainable competitive advantage, even to larger non-family firms (Chang et al., 2022), through their long-term orientation to the environment (Bichler et al., 2021), deep community ties, and strong value system that keeps the socioemotional wealth (SEW) alive as long as the firm exists.

Long-term perspective. The long-term perspective of competitive advantage is the aim of every firm (Barringer & Ireland, 2019; Kotler et al., 2020b; Rothaermel, 2017; Tueth, 2010). Naturally, family firms look beyond the short-term gains unlike non-family firms or publicly shared firms. In such cases, shareholders want to be better off in the short term, while family-owned companies look for long-term sustainability built on competitive advantage.

Strong social ties. Corporate social responsibility (CSR) is one of the postulates of how family firms are embedded in society. CSR represents a valuable means for the family firms to develop the brand name in the society (Chiang et al., 2022; Lumpkin & Bacq, 2022), and it works in most cases. Yet, family firms are more open to making a positive and extraordinary relationship with the community, the local and foreign firms (Dana et al., 2000) in the region they operate. They contribute in different meaningful ways to help the society to resolve current issues or any ad hoc problem that needs a sponsor or an investor.

Family firms use these opportunities very thoughtfully (Dayan et al., 2019) because they look for the long-term embeddedness in the society, and their own stability (Fernandez, 2023). For instance, if the society needs to resolve some environmental issues, they first ask family firms for help. Usually, the family enterprise responds positively, which brings to the enterprise additional values that will be reflected in the long-term perspective.

Strong-inherited value system. The most important value in family firms is socioemotional wealth (Barros et al., 2016, 2017; Cleary et al., 2019; Debicki et al., 2016; Gómez-Mejía et al., 2007; Gómez-Mejía & Herrero, 2022; Gomez-Mejia et al., 2023; Kalm & Gomez-Mejia, 2016).

SEW is used to promote family and business values at the same time because reputation of the family businesses is always related to family's name. Therefore, a family firm's reputation plays an important role in creating, attaining, and sustaining competitive advantage in the long term. Yet, environmentally safe products provided by the family business will enhance the family reputation and it will make it stronger in the community, society, and the region the family firm operates.

2.6 Implications of Triple-Bottom Approach for Sustainable Competitive Advantage

As argued by Rothaermel (2017), the competitive advantage is "relative" because if it is not sustained, then it can be diminished over time, yet quickly. Therefore, firms are trying to navigate the best routes that will lead them toward long-term sustainability.

Family firms are the ones that also seek a sustainable competitive advantage. Many authors argue that family businesses have a lot of savvy (Dayan et al., 2019) so that they wisely use environmental opportunities to sustain their leadership in the industry and the market. Moreover, nowadays, they will use digital opportunities which will leverage their position in the market. Similarly, consumers will stay loyal and support their brand in the market.

Non-financial value (SEW) that distinguishes family firms from non-family ones, plays an important role in the sustainable strategies of the family firms. Some perceive that the SEW prevents future investments and may prevent the development of the family firm's competence and capabilities (Chrisman & Patel, 2012; Delgado-García et al., 2023), however, some have opposite arguments in which SEW actually builds the image in the society and in the environment (Lumpkin & Bacq, 2022), so that helps the family firms gain its sustainability.

As the new agenda has called for triple-bottom-line approach to save the planet, to foster socio-economic development, and being awarded for being engaged in it (by profit), family firms have positioned themselves at a very high level of it. They have very strong social ties (Palalić et al., 2019; Palalić & Smajić, 2022; Smajić et al., 2022) which qualifies them to be foremost enterprises to take the lead in CSR projects in their community.

Finally, family firms gain their sustainable competitive advantage via implementation of the TBL approach (Fig. 2.2) in a way that they monitor the environmental

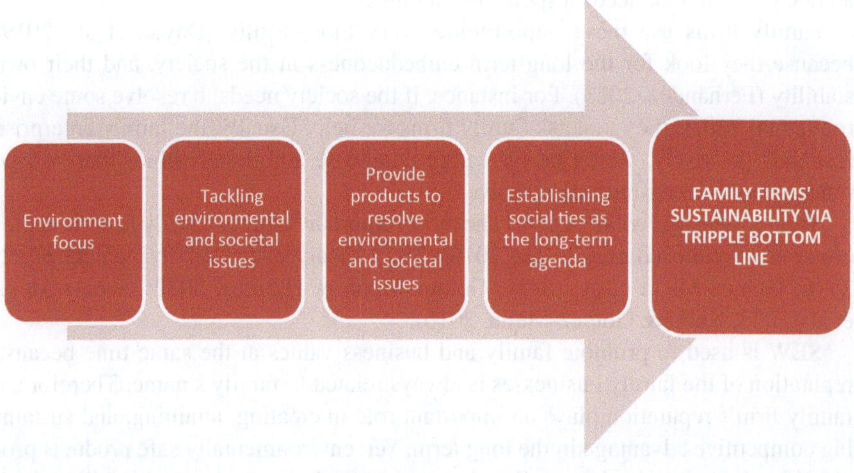

Fig. 2.2 Pathway toward long-term sustainability using triple-bottom-line approach. Source: Authors

and societal changes and act accordingly. Moreover, in monitoring the changes in the environment, family enterprises act proactively to solve the environmental and societal issues while satisfying customers and gaining value in return(profit). Having said this, family firms mindfully and thoughtfully prepare themselves and their successor to advance the current operations.

Meantime, they build strong social ties (Zellweger et al., 2012) to scale their operations locally. If appropriate, they expand their operations globally (as discussed in Chap. 8).

Case Study[1]: *Microsoft*
"Be what's next"

Forty-seven years ago on April 4th, 1975, at a phase when the majority of people were using typewriters, Bill Gates and Paul Allen were working toward putting the last block of Microsoft company. Today in 2022, Microsoft company is considered the world's biggest computing hardware and software multinational corporation. It has more than 101 diverse products, apps, and services for its professional and commercial consumers.

Microsoft's sustainability and long-lasting success is due to its competitive advantage. Its dominant position in the computing sector stems from a range of significant strengths that form its competitive edge over its biggest competitors which are Apple, Amazon, and Google. Microsoft's continuous emphasis over innovation has assisted it to maintain its strong position in the intensely competitive industry. Through innovating, Microsoft aims to provide superior solutions in terms of products and services, that assist people in becoming more productive. Furthermore, by comprehensively studying the market, it acquires businesses that will boost its value. For Instance, in 2016 it acquired LinkedIn, with the objective of strengthening the professional networking site and integrating it with its corporate tools like Office 365. In addition, it has also penetrated gaming market, by releasing Xbox in 2001, and it captured a considerable market share. Moreover, its competitive advantage is further realized by developing an optimal pricing strategy, gaining customer loyalty, having a strong brand equity, and being globally present. Besides, having a good reputation for its clear commitment to its Social and Environmental responsibilities, along with superior financial performance.

It is worthy mentioning that Microsoft has the highest ESG score compared to the other Dow Jones-listed companies, which equals 76.30 out of 100. The ESG investing approach stands for Environmental, Social, and Corporate Governance, and it concentrates on companies that are committed to environmental preservation, social responsibility, and ethical corporate practices.

High ESG-scored companies provide uppermost chance of long-standing success for the investors; because the ranking provided by the ESG criteria incorporates the three main non-financial factors, besides profits. Furthermore, ESG gauges the company's ethical effect and sustainability, thus, the shares of companies with high ESG rankings are identified as sustainable assets and investors are attracted to invest in them. The Environmental and social aspects of ESG are commonly shared with the Triple-bottom-line (TBL) approach.

TBL is an approach introduced in 1994 by John Elkington and it lies in the notion of "Doing Well by Doing Good." Elkington states that if a company wants to grow and capitalize over a large market share, then it should not only focus on its financial performance but also focus on its impact on the society and the planet. Microsoft's achievement of a high ESG score ascertains its positive impact on the society and environment, along with its superior financial performance. Microsoft's creation of "Bill & Melinda Gates foundation," further emphasizes its social roles. It aims to assist people globally in attaining a healthy, financially stable, and productive life.

In addition, Microsoft has offered $2.5 billion as technology grants, and discounts to 295,000 non-profit parties in 2021. Besides, offering the "cloud-first grant program," that allows the non-profits to gain access to world-class cloud technology globally, hence accomplishing their tasks effectively and rapidly. Moreover, it has launched "AccountGuard" program for non-profits to address the rising threat of cyberattacks at no additional expense.

Moving to its environmental roles, it plans to become carbon negative by 2030, after being carbon neutral since 2012. Therefore, in 2015, it has acquired over 3.2 billion KW hours of zero-carbon energy and engaged with its suppliers and utilities to enhance renewable energy access and cost. Moreover, it creates tools that solve water-related issues globally and it is keen to minimize its use of water and to conserve it within its operational areas.

Furthermore, it aims to lower and eliminate waste at its operations by reusing its resources, source reduction, and recycling. Microsoft's efforts to meet its environmental and social responsibilities have not stopped it from achieving a strong financial position, but further assisted it to be the world's third most valuable firm in 2022 with a market capitalization of $2.055 Trillion.

Furthermore, Microsoft has planned and conducted practices on the departmental level to further ensure the optimal implementation of Triple-Bottom-Line (TBL) approach, to have a superior effect on society and planet. For instance, the Human Resources Management (HRM) Department's team, in particular, has demonstrated evident efforts in conducting their tasks and setting rules to ensure that the company's social and environmental responsibilities are properly met.

Starting with HRM's efforts in meeting its social responsibilities, it published a Global Human Rights statement in 2012. The statement addresses the appropriate practices to be followed when interacting with suppliers, customers, employees, and the community as a whole, in order to preserve human rights and have a positive impact on people.

Moreover, the statement explicitly tackles the company's commitment to the abolition of all forms of discrimination against women and supporting the rights of children along with the rights of people with disabilities. In addition, Microsoft's employees are trained to appropriately apply the practices to ensure the proper fulfillment of social responsibilities. Furthermore, the HRM program of "Microsoft Employee Giving," encourages the employees to volunteer their time, skill, and money to the non-profits along with training them to optimally utilize the donated tools and technologies of Microsoft.

Moving to the environmental influence of the company under HRM, they undertake waste reduction strategies such as office equipment reuse, recycling, and composting. In addition, they have established an auditing procedure to ensure that the assigned firm for processing the waste, recycles or disposes of it ethically and in accordance with environmental standards. More importantly, Microsoft recognizes the negative impact of the electronic waste (e-waste) generated by its internal activities.

Therefore, it has developed the IT Asset Disposition Program (ITAD) program, which promotes the reuse and recycling of e-waste, reducing life cycle energy usage and greenhouse gas emanations along with reducing harmful waste.

Regarding the financial aspect, HRM team carefully chooses the candidates for the available positions, depending on specific skills and knowledge required. Through implementing practices that will enhance society and protect the environment, Microsoft has succeeded in becoming one of the leading multinational corporations that illustrate social and environmental responsibility along with achieving a superior financial position. Additionally, it is classified as the highest-ranked company by ESG criteria.

Questions for discussion:

1. Do you think that Microsoft does the right thing(s)? Why? Why not?
2. Based on the information above, can one affirm the sustainable competitive advantage of Microsoft, the way it does?
3. Do you think that Microsoft truly maximizes the triple-bottom line to achieve its long-term competitive advantage (sustainability)?
4. Is it enough what Microsoft does for its sustainability?
5. What else you would suggest as a paramount action to achieve and sustain its competitive advantage via the TBL approach?

Bibliography

A&E Television Networks. (2020, April 1). *Microsoft founded*. History. Retrieved April 26, 2022, from https://www.history.com/this-day-in-history/microsoft-founded

Cowling, J. (2016, February 8). *A brief history of Microsoft – the worlds biggest software company*. DSP explorer. Retrieved April 26, 2022, from https://content.dsp.co.uk/a-brief-history-of-microsoft-the-worlds-biggest-software-company

DePersio, G. (2022, November 3). 3 trends to watch in ESG investing. Investopedia. Retrieved April 26, 2022, from https://www.investopedia.com/articles/investing/030316/3-trends-watch-esg-investing.asp

EMBAPro. (n.d.). Microsoft (Germany) triple bottom line analysis. embapro. Retrieved April 26, 2022, from https://embapro.com/frontpage/tblcoanalysis/7670-microsoft

Environmental sustainability: Microsoft CSR. Microsoft. (n.d.). Retrieved April 26, 2022, from https://www.microsoft.com/en-us/corporate-responsibility/sustainability?activetab=pivot_1%3Aprimaryr3

Erlichman, J. (2019, October 23). Three years after Microsoft acquisition, LinkedIn keeps quietly climbing. BNN Bloomberg. Retrieved April 26, 2022, from https://www.bnnbloomberg.ca/three-years-after-microsoft-acquisition-linkedin-keeps-quietly-climbing-1.1335990

Foundation fact sheet. Bill & Melinda Gates Foundation. (n.d.). Retrieved April 26, 2022, from https://www.gatesfoundation.org/about/foundation-fact-sheet

Microsoft (MSFT) – market capitalization. CompaniesMarketCap.com. (2022). Retrieved April 26, 2022, from https://companiesmarketcap.com/microsoft/marketcap/

Microsoft 2016 Corporate Social Responsibility. Microsoft. (2016). Retrieved April 30, 2022, from https://www.microsoft.com/en-us/CMSFiles/Microsoft_2016_Corporate_Social_Responsibility.pdf?version=889768cf-2300-6a48-33e1-5fca73a1836e&CollectionId=df8dab12-dbf6-441f-a2db-5996225f2c6a

Microsoft buys LinkedIn. Microsoft-Stories. (2020, September 3). Retrieved April 26, 2022, from https://news.microsoft.com/announcement/microsoft-buys-linkedin/

Microsoft Global Human Rights. Human rights statement | Microsoft CSR. (n.d.). Retrieved April 30, 2022, from https://www.microsoft.com/en-us/corporate-responsibility/human-rights-statement?activetab=pivot_1%3Aprimaryr8

Mohan, M. (2021, January 10). Over 101 Microsoft Products, services, & apps you probably don't know. Minterest. Retrieved April 26, 2022, from https://www.matrics360.com/microsoft-products-and-services/

Pratap, A. (2018, December 19). Microsoft sources of competitive advantage. notesmatic. Retrieved April 26, 2022, from https://notesmatic.com/microsoft-sources-of-competitive-advantage/

Spelhaug, J. (2021, October 4). Microsoft Tech for Social Impact announces its updated cloud-first grant program. Microsoft. Retrieved April 26, 2022, from https://blogs.microsoft.com/on-the-issues/2021/10/04/microsoft-tech-for-social-impact-cloud-grant-program/

Staff, I. (2021, November 10). 100 best ESG companies: Top stocks for environmental, social and governance values. Investor's Business Daily. Retrieved April 26, 2022, from https://www.investors.com/news/esg-companies-list-best-esg-stocks-environmental-social-governance-values/
Stopps, N. (2021, January 27). Why ESG ratings matter and how companies use them. Simply Sustainable. Retrieved April 26, 2022, from https://simply-sustainable.co.uk/insights/why-esg-ratings-matter-and-how-companies-use-them

Questions for Discussion

- Define the competitive advantage.
- Define sustainable competitive advantage.
- Discuss and explain the importance of sustainable competitive advantage. Provide appropriate examples.
- Discuss tools and mechanisms of sustainable competitive advantage in family firms.
- Explain how cost leadership strategy affects competitive advantage of family businesses.
- Explain how differentiation strategy affects competitive advantage of family businesses.

Additional Readings

- Yin, Y., Crowley, F., Doran, J., Du, J. and O'Connor, M. (2023), Research and innovation and the role of competition in family owned and managed firms. *International Journal of Entrepreneurial Behavior & Research,* Vol. 29 No. 1, pp. 166–194. https://doi.org/10.1108/IJEBR-12-2021-1031
- Ramadani, V. & Hoy, F. (2015), Context and uniqueness of family businesses. In Dana, L-P. & Ramadani, V. (Eds.), Family Businesses in Transition Economies: Management, Succession and Internationalization (pp 9–37), Cham: Springer.
- Schweiger, N., Matzler, K., Hautz, J., & de Massis, A. (2023). Family businesses and strategic change: the role of family ownership. *Review of Managerial Science*. https://doi.org/10.1007/s11846-023-00703-3
- Hall, A., Melin, L., & Nordqvist, M. (2001) Entrepreneurship as radical change in the family business: Exploring the role of cultural patterns. *Family Business Review*, 14(3), 193–208.
- Kalm, M., & Gomez Mejia, L.R. (2016). Socioemotional wealth preservation in family firms, Revista de Administração 51, 409–411.
- Krappe A, Goutas L, von Schlippe A (2011) The "family business brand": an enquiry into the construction of the image of family businesses. *Journal of Family Business Management* 1: 37–46.

- Micelotta & Raynard, (2011). Micelotta, E., & Raynard, M. (2011). Concealing or revealing the family? Corporate brand identity strategies in family firms. *Family Business Review,* 24, 197–216.

Suggested Activities

- Find two family businesses from different regions. Identify the basis of their competitive advantage.
- In the above-selected family firms, try to identify how much of the TBL approach is implemented by each family business.
- Select two to three family firms in your neighborhoods. Investigate each firm's image status in the society. How much it has effect on the environmental engagement of family firms?
- Identify the top 5 family firms in your country and compare their involvement in the TBL approach. Identify their weaknesses in implementing the TBL approach.

Summary of Key Concepts and Keywords

- Competitive advantage
- Sustainable competitive advantage
- Differentiation strategy
- Cost leadership strategy
- CSR
- Societal concept
- Economic concept
- Environmental concept

References

Websites

Ahmad, N., Smajić, H., Palalic, R., & Al-Riyami, S. (2022). Sustainability management research in the Arab world: A citation-based systematic literature review. *Management & Sustainability: An Arab Review, 1*(1), 14–37. https://doi.org/10.1108/msar-01-2022-0004

Barringer, B. R., & Ireland, R. D. (2019). *Entrepreneurship: Successfully launching new ventures* (6th ed.). Pearson.

Barros, I., Hernangómez, J., & Martin-Cruz, N. (2016). A theoretical model of strategic management of family firms. A dynamic capabilities approach. *Journal of Family Business Strategy, 7*(3), 149–159. https://doi.org/10.1016/j.jfbs.2016.06.002

Barros, I., Hernangómez, J., & Martin-Cruz, N. (2017). Familiness and socioemotional wealth in Spanish family firms: An empirical examination. *European Journal of Family Business, 7*(1–2), 14–24.

Bichler, B. F., Kallmuenzer, A., Peters, M., Petry, T., & Clauss, T. (2021). Regional entrepreneurial ecosystems: How family firm embeddedness triggers ecosystem development. *Review of Managerial Science, 16*(1), 15–44. https://doi.org/10.1007/s11846-020-00434-9

Chang, E. P. C., Zare, S., & Ramadani, V. (2022). How a larger family business is different from a non-family one? *Journal of Business Research, 139*, 292–302. https://doi.org/10.1016/j.jbusres.2021.09.060

Chiang, H.-T., Lin, Y.-C., & Chen, W.-W. (2022). Does family business affect the relationship between corporate social responsibility and brand value? A study in different industry Taiwan. *Asia Pacific Management Review, 27*(1), 28–39. https://doi.org/10.1016/j.apmrv.2021.04.002

Chrisman, J. J., & Patel, P. C. (2012). Variations in R & D Investments of Family and Nonfamily Firms: Behavioral agency and myopic loss aversion perspectives. *Academy of Management Journal, 55*(4), 976–997. https://doi.org/10.5465/amj.2011.0211

Cleary, P., Quinn, M., & Moreno, A. (2019). Socioemotional wealth in family firms: A longitudinal content analysis of corporate disclosures. *Journal of Family Business Strategy, 10*(2), 119–132. https://doi.org/10.1016/j.jfbs.2018.11.002

Dana, L. P., Etemad, H., & Wright, R. W. (2000). The global reach of symbiotic networks. *Journal of Euro Marketing, 9*(2), 1–16. https://doi.org/10.1300/J037v09n02_01

Dayan, M., Ng, P. Y., & Ndubisi, N. O. (2019). Mindfulness, socioemotional wealth, and environmental strategy of family businesses. *Business Strategy and the Environment, 28*(3), 466–481.

Debicki, B. J., Kellermanns, F. W., Chrisman, J. J., Pearson, A. W., & Spencer, B. A. (2016). Development of a socioemotional wealth importance (SEWi) scale for family firm research. *Journal of Family Business Strategy, 7*(1), 47–57. https://doi.org/10.1016/j.jfbs.2016.01.002

Dehning, B., & Stratopoulos, T. (2003). Determinants of a sustainable competitive advantage due to an IT-enabled strategy. *The Journal of Strategic Information Systems, 12*(1), 7–28. https://doi.org/10.1016/s0963-8687(02)00035-5

Delgado-García, J. B., Blanco-Mazagatos, V., Romero-Merino, M. E., & Díaz-Portugal, C. (2023). Family CEO affect and R & D investments of family firms: The moderation effect of family ownership structure. *Long Range Planning, 56*(5), 102230. https://doi.org/10.1016/j.lrp.2022.102230

Fernandez, V. (2023). Family entrepreneurship around the world. *International Review of Financial Analysis, 89*, 102808. https://doi.org/10.1016/j.irfa.2023.102808

Gómez-Mejía, L. R., Haynes, K. T., Núñez-Nickel, M., Jacobson, K. J. L., & Moyano-Fuentes, J. (2007). Socioemotional wealth and business risks in family-controlled firms: Evidence from Spanish olive oil Mills. *Administrative Science Quarterly, 52*(1), 106–137. https://doi.org/10.2189/asqu.52.1.106

Gómez-Mejía, L. R., & Herrero, I. (2022). Back to square one: The measurement of socioemotional wealth (SEW). *Journal of Family Business Strategy, 13*(4), 100480. https://doi.org/10.1016/j.jfbs.2021.100480

Gomez-Mejia, L. R., Mendoza-Lopez, A., Cruz, C., Duran, P., & Aguinis, H. (2023). Socioemotional wealth in volatile, uncertain, complex, and ambiguous contexts: The case of family firms in Latin America and the Caribbean. *Journal of Family Business Strategy, 15*(1), 100551. https://doi.org/10.1016/j.jfbs.2022.100551

Kalm, M., & Gomez-Mejia, L. R. (2016). Socioemotional wealth preservation in family firms. *Revista de Administração, 51*(4), 409–411. https://doi.org/10.1016/j.rausp.2016.08.002

Knudsen, E. S., Lien, L. B., Timmermans, B., Belik, I., & Pandey, S. (2021). Stability in turbulent times? The effect of digitalization on the sustainability of competitive advantage. *Journal of Business Research, 128*, 360–369. https://doi.org/10.1016/j.jbusres.2021.02.008

Kotler, P., Armstrong, G., Harris, L. C., & He, H. (2020a). *Principles of marketing*. Pearson.

Kotler, P., Armstrong, G., Harris, L. C., & Hongwei, H. (2020b). *Principles of marketing* (8th ed.). Pearson.

Lumkin, G. T., Scott, B. D., & Dess, G. G. (2002). Achieving sustainable competitive advantage and avoiding pitfall. *Organizational Dynamics, 30*(4), 325–340.

Lumpkin, G. T., & Bacq, S. (2022). Family business, community embeddedness, and civic wealth creation. *Journal of Family Business Strategy, 13*(2), 100469. https://doi.org/10.1016/j.jfbs.2021.100469

Palalić, R., Ramadani, V., Welsh, D. H. B., Dizdarević, A., & Đilović, A. (2019). Impact of networking and business environment on student leadership styles. *International Entrepreneurship and Management Journal, 17*(1), 447–475. https://doi.org/10.1007/s11365-019-00610-8

Palalić, R., & Smajić, H. (2022). Socioemotional wealth (SEW) as the driver of business performance in family businesses in Bosnia and Herzegovina: The mediating role of transformational leadership. *Journal of Family Business Management, 12*(4), 1043–1064. https://doi.org/10.1108/jfbm-07-2021-0067

Ramadani, V., Memili, E., Palalić, R., & Chang, E. P. C. (2020). *Entrepreneurial family businesses*. Springer. https://doi.org/10.1007/978-3-030-47778-3

Rothaermel, F. (2017). *Strategic management* (3rd ed.). Mc Graw Hill Education.

Rothaermel, F. T. (2015). *Strategic management* (2nd ed.). McGraw-Hill Education.

Smajić, H., Palalić, R., & Ahmad, N. (2022). Future perspective of socioemotional wealth (SEW) in family businesses. *Journal of Family Business Management, 13*(4), 923–954. https://doi.org/10.1108/jfbm-05-2022-0070

Tueth, M. (2010). *Fundamentals of sustainable business: A guide for the next 100 years*. World Scientific.

Zellweger, T. M., Kellermanns, F. W., Eddleston, K. A., & Memili, E. (2012). Building a family firm image: How family firms capitalize on their family ties. *Journal of Family Business Strategy, 3*, 239–250.

Sustainable Finance in Family Businesses

3

Learning Objectives
After reading this chapter, you should be able to:

- Define sustainable finance.
- Understand the SDG goals.
- Know the role of sustainable finance in family businesses.
- Know the framework for sustainable finance.
- Understand the sustainable finance categories.
- Know the benefits of sustainable finance.
- Know the challenges of sustainable finance.

Profile: Transocean, Ltd

Steven Newman served as the Chief Executive Officer at Transocean, Ltd., a prominent offshore drilling contractor employing over 18,000 individuals globally. His background lies in petroleum engineering, who have dedicated significant time to working on drilling rigs, accumulating over 15 years of experience at Transocean, Ltd. He took pride in the company's historical advancements in the energy sector during my tenure.

On April 20, 2010, the Deepwater Horizon drilling platform, owned by Transocean, Ltd. and operated by British Petroleum in the Macondo Prospect in the Gulf of Mexico, experienced a catastrophic oil spill. The incident was triggered by an explosion that claimed the lives of 11 workers and resulted in the largest accidental marine oil spill in the history of the petroleum industry. The United States Government estimated the total discharge to be 4.9 million barrels, leading to a relentless effort to contain the spill. Mr. Newman in his

R. Palalić et al., *Sustainable Family Business*, Springer Texts in Business and Economics, https://doi.org/10.1007/978-3-031-57999-8_3

testimony in front of the Committee on Energy & Natural Resources, United States Senate, On May 11, 2010, said "The last few weeks have been a time of great sadness and reflection for our Company - and for me personally. Nothing is more important to me and Transocean than the safety of our employees and crew members, and our hearts ache for the widows, parents, and children of the 11 crew members - including nine Transocean employees - who died in the Deepwater Horizon explosion."

Despite numerous attempts to control the flow, the well was eventually declared sealed on September 19, 2010. A massive response was initiated to safeguard beaches, wetlands, and estuaries, employing skimmer ships, floating booms, controlled burns, and oil dispersants. Cleanup efforts extended along 55 miles of the Louisiana shoreline until 2013, and traces of oil were discovered as far as the waters of the Florida Panhandle and Tampa Bay, with the oil and dispersant mixture embedded in the sand.

The prolonged spill, coupled with the adverse impacts of response and cleanup activities, inflicted extensive damage on marine and wildlife habitats and the fishing and tourism industries. Numerous investigations delved into the causes of the explosion and record-breaking spill.

The US government's September 2011 report notably attributed the disaster to defective cement on the well, primarily assigning blame to BP. However, it also implicated rig operator Transocean and contractor Halliburton. In an earlier assessment in 2011, a National Commission held BP and its partners accountable for cost-cutting decisions and an inadequate safety system. The commission concluded that the spill resulted from "systemic" root causes, warning that without "significant reform in both industry practices and government policies, might recur."

Source: Based on Schoenmaker (2017).

3.1 Introduction

In a world with limited resources, the management of both personal and professional aspects of life has been significantly influenced. This shift is not exclusive to individuals but extends to businesses, which have embraced innovative approaches to conduct operations, aiming to conserve resources for the benefit of future generations. This evolving landscape has prompted business activities and operations changes over the years (Kajtazi et al., 2023). Historically, businesses operated using traditional methods. However, the current scenario has witnessed a transformative shift, where the adoption of green finance principles aligns with the broader objective of resource conservation and sustainable practices. In contrast to the conventional model, the integration of green finance in family businesses represents a departure from traditional business practices toward a more environmentally conscious and innovative approach. Just as information technology revolutionized the broader business landscape, the incorporation of green finance in family businesses marks a shift from the conventional to a more sustainable and forward-thinking model, focusing on economic, and environmental dimensions (Rexhepi et al., 2013). This transition reflects a commitment to not only optimizing present business processes but also ensuring that these enterprises play a role in preserving resources for the well-being of future generations (Kajtazi et al., 2023).

Sustainable finance tends to bring these issues closer to the organization. According to the European Commission, sustainable finance is defined as the process of taking due account of environmental, social, and governance (ESG) considerations when making investment decisions in the financial sector, leading to increased longer-term investments into sustainable economic activities and projects. Sustainable finance also tends to orient family businesses to think on three dimensions social, economic, and environmental dimensions. Social Dimension of sustainable finance which often are neglected by many organizations, it underscores the significance of social indicators in both business and social sectors. This encompasses stakeholder management, socially responsible investing, social inclusion and equality, labor and human rights, community development, philanthropy, and transparency of social performance (Sepasi et al., 2018). While the economic dimension is related to addressing issues related to jobs, ethical training standards, and product value, the economic dimension is linked to creating employment opportunities, delivering higher-quality products and services, improving standards in countries where entrepreneurs operate, and boosting revenues in the state budget. Environmental dimension on the other hand is prominently highlighted in organizations with a sustainability focus. It involves environmental management (including systems, water–energy efficiency, services, and supply chain), natural preservation (such as climate change mitigation efforts, biodiversity, waste, and emission reduction), and general categories (like risk assessment, transparency of environmental performance, and participation in nature conservation efforts) (Rexhepi et al., 2018; Sepasi et al., 2018).

The dimensions of sustainable finance are related to Economic, Social, and Governance (ESG) criteria presented by the European Commission in 2015, known as 17 SDGs (see Chap. 1).

In the context of family businesses, the application of sustainable finance aligns closely with these dimensions, as it represents a strategic approach to integrating environmental responsibility into financial decision-making. The success of sustainability finance is primarily contingent on the entrepreneur who serves as the founder of the organization. Crafted to benefit entire economies, sustainability finance has garnered significant attention, becoming a frequent topic of discussion in various forums, including conferences, seminars, research journals, universities, government entities, municipalities, and publishers. According to a survey conducted by JP Morgan and the Global Impact Investing Network (GIIN) in November 2010, the social finance market was estimated to range from $183 billion to $667 billion, with invested capital ranging from $400 billion to nearly $1 trillion (O'Donohoe et al., 2010). A study by the Global Impact Investment Network and JP Morgan forecasts a nearly 20% growth in the amount invested by the 125 leading impact investors in social finance this year (European Commission, 2015). Notably, G8 countries, particularly the United Kingdom and the United States, have taken initiatives to create new social investment models, with similar interests and activities emerging globally. These initiatives, spearheaded by governments, foundations, investors, and other stakeholders, have played a pivotal role in accelerating sustainable finance market in recent years (Wilson, 2014; Rexhepi, 2016).

Within the spectrum of investment categories, sustainable finance encompasses aspects of both philanthropy and socially responsible investing, amalgamating their features. Analogous to philanthropy, sustainable finance organizations prioritize societal well-being over the pursuit of maximum profits, channeling primary earnings to employees and investors, frequently citizens. In alignment with socially responsible investing, these entities deliberately forego profit maximization, opting instead to provide affordable, high-quality products and employ individuals facing challenges, including the homeless and disabled. They consistently address societal issues and actively uphold environmental considerations (Rexhepi, 2016).

This heightened awareness has led to increased public concern and stakeholder interest in the natural environment, prompting a transformation of the competitive landscape, which has pushed organizations to adopt more green innovation strategies in their strategies (Rexhepi et al., 2023). In this context, the emergence of sustainable finance has become a novel strategic orientation for organizations. Although a relatively recent concept, sustainable finance has been extensively researched over the past two decades from various perspectives Rexhepi et al., 2023). Sustainability, as a ubiquitous term, encompasses a spectrum ranging from preserving the planetary natural environment to ensuring long-term profitability through a competitive position (Rahdari et al., 2020). The adoption of sustainable finance in family businesses emerges as a strategic response to these multifaceted challenges, emphasizing the need to balance economic objectives with environmental responsibility for the benefit of current and future generations Rexhepi et al., 2023). Even though sustainable finance has gained more importance where 82% of organizations worldwide intend to increase their ESG investments, however, this is still limited and more needs to be done (Zamfiroiu & Pînzaru, 2021). The current data available on sustainable finance is still scarce in many countries, however, this has changed through the years. Consequently, having more data it will embrace the trend of sustainable finance and investing and increase the awareness for more long-term versus medium and short-term perspectives (Zamfiroiu & Pînzaru, 2021).

3.2 Family Business and Sustainable Finance

Bill and Melinda Gates occupy a prominent position within the global philanthropic sphere, garnering widespread acclaim for their substantial charitable contributions. In the year 2012 alone, they directed an astounding sum of 3.4 billion dollars toward philanthropic causes. Publicly affirming their dedication to sustaining this philanthropic ethos, the Gates committed a significant portion of their considerable wealth for the foreseeable future. Similarly, Warren Buffet, a notable billionaire and investor, has embarked on a noteworthy philanthropic journey, pledging to donate a substantial 99% of his amassed wealth. This admirable commitment extends beyond the Gates family, as numerous billionaires, affluent individuals, and governmental entities contribute to diverse philanthropic endeavors. While this collective generosity is commendable, it prompts a pertinent question: Can such benevolent acts effectively address the pervasive issue of global poverty? (Rexhepi et al., 2013).

To delve into this inquiry, one may contemplate a hypothetical scenario wherein all billionaires choose to allocate 80% of their wealth to charitable causes. This theoretical wealth redistribution would primarily funnel resources through charitable organizations, which would then allocate assistance to those in need, thus mitigating poverty for a segment of the population, albeit temporarily. However, the sustainability of this approach in permanently eradicating world poverty remains uncertain, primarily due to concerns regarding the allocation and utilization of these funds. Sustainability is the best solution to many issues presented in this book and Family businesses need to play an important role in this process since they are the backbone of economic growth in both industrialized and developing countries, they constitute greater than 80% of all businesses in the world and contribute 70–90% of the global GDP, creating more than half of the total employment in the developed countries (Oudah et al., 2018). Recently, sustainable finance in family businesses has gained significant traction, becoming a recurring theme in academic conferences, seminars, research journals, universities, governmental bodies, municipalities, and the publishing sector. This shift presents a promising framework for addressing global poverty in a more enduring and holistic manner Rexhepi et al., 2023). The concept of "green" and "sustainable finance," as conceptualized by the European Commission, is designed to achieve several crucial objectives.

- ***Firstly***, it aims to mobilize increased investment from the European Union and the global community, facilitating the generation, attraction, circulation, and transfer of capital within Europe. This capital is intended to support Small and Medium Enterprises (SMEs), infrastructure projects, and long-term sustainable initiatives, contributing to solutions that address households' retirement goals, expand business opportunities, foster economic growth, create jobs, and promote inclusive and robust sustainable development (European Commission, 2015, p. 3).
- ***Secondly,*** the initiative seeks to enhance the connection between financing and investment projects across the EU, promoting the development of small markets and meeting the growth needs of businesses and projects (European Commission, 2015, p. 3).
- ***Thirdly,*** it aims to establish a more stable financial system by leveraging integrated financial and capital markets to enable Member States, especially those in the euro area, to collectively manage the impact of economic shocks. This approach helps share financial risks, making EU citizens and companies less vulnerable to banking contractions and encourages investments in more devel-

oped equity markets rather than increased indebtedness (European Commission, 2015, p. 3).

- ***Lastly,*** the initiative focuses on deepening financial integration and intensifying competition, anticipating benefits such as increased cross-border risk-sharing, deeper and more liquid markets, and diversified funding sources, ultimately fostering greater European competitiveness (European Commission, 2015, p. 3).

The integration of sustainable finance into family businesses holds profound implications. Firstly, it contributes to environmental sustainability by mitigating the environmental impact of family operations and investments. Secondly, it aligns familial values with financial objectives, fostering a sense of purpose and a shared commitment among family members (Murray & Williard, 2018). Thirdly, Sustainable Finance can result in cost savings and heightened efficiency through the adoption of eco-friendly technologies and sustainable practices (Lund & Myers, 2020). Moreover, it has the potential to enhance the reputation of the family business, attracting environmentally conscious customers, partners, and investors, thereby improving market positioning.

At its essence, sustainable finance involves making financial decisions that consider the broader environmental, social, and governance (ESG) impacts of investments and operations. This approach means that family businesses need to integrate ESG factors into financial planning, risk management, and investment strategies and focus beyond the current generation, embracing a sense of responsibility for the legacy of future generations (Gomez & Anderson, 2018a). They prioritize stability and long-term viability, key tenets of sustainable finance that aim to balance financial returns with social and environmental objectives (Brown & Miller, 2017). This approach is a critical component of the broader sustainability agenda and is increasingly gaining prominence in academic, corporate, and policy discussions (Bauer & Hann, 2018).

However, despite the increasing recognition of sustainable finance, it remains a relatively novel concept, not universally well understood. Research indicates that sustainable finance tends to be more familiar to individuals engaged in specialized financial sectors such as socially responsible investing and credit unions. This chapter aims to elucidate the concept of sustainable finance and its underlying framework by providing a comprehensive explanation and highlighting the prospective societal benefits of sustainable finance in the future.

3.2.1 Principles of Sustainable Finance in Family Business

As previously elucidated, a fundamental tenet of sustainable financing involves the incorporation of Environmental, Social, and Governance (ESG) criteria into investment decisions, as highlighted by Bauer and Hann (2018). Investors and financial institutions no longer assess investments solely on their potential financial returns;

rather, they scrutinize them for their environmental impact, social implications, and governance practices. The adoption of essential sustainability principles is imperative for the successful implementation of sustainable finance in family businesses, as asserted by Davis et al. (2020). Several fundamental principles underpin sustainable finance in family businesses:

- **Intergenerational Equity:** An essential tenet of sustainable finance in family businesses is intergenerational equity. Practitioners of this principle seek to balance the needs and aspirations of the current generation with those of future generations. This involves preserving and growing family wealth while simultaneously safeguarding environmental and societal well-being for heirs yet to come.
- **Alignment with Family Values:** Sustainable finance must align with the core values of both the family and the business. By incorporating these values into financial decision-making, family businesses can ensure that their investments and operations align with their overarching mission and ethos. The foundation of sustainable finance in family businesses lies in integrating financial activities with the family's core values, emphasizing the importance of incorporating Environmental, Social, and Governance (ESG) objectives consistent with familial beliefs and ethos (Basak, 2019).
- **Long-Term Vision:** Sustainable finance advocates for a long-term vision over short-term gains. This perspective encourages family businesses to adopt strategies that not only generate immediate profits but also create lasting value over the years. Investments in sustainable technologies, social responsibility, and environmental stewardship exemplify this principle. Sustainable finance in family businesses underscores the importance of long-term orientation, emphasizing that financial decisions should consider their lasting impact on the family, the business, and the environment rather than focusing solely on short-term gains (Murray & Williard, 2018).
- **Transparency and Accountability:** Transparency and accountability are fundamental principles (Davis et al., 2020). Family businesses must openly disclose their sustainability objectives, report on their performance, and be accountable for the impact they generate. This fosters trust among stakeholders and ensures commitment to ESG goals.
- **Responsible Investment Strategies:** Funds should be directed toward investments and ventures that showcase a commitment to environmental and social responsibility, encompassing areas like green technologies, equitable labor practices, and community development initiatives. Sustainable finance mandates the adoption of conscientious investment approaches (Gomez & Anderson, 2018b).
- **Risk Management:** Family enterprises ought to evaluate and address financial risks linked to climate change, social challenges, and ethical considerations, guaranteeing their adaptability in addressing the ever-changing landscape of sustainability challenges.

Implementing these principles in family businesses may face challenges. These include tensions between short-term financial objectives and long-term sustainability goals, resource and expertise constraints, and the potential impact of succession planning on the continuity of sustainable finance practices (Davis et al., 2020). Overcoming these challenges requires a thoughtful and tailored approach, often involving the development of clear governance structures, open communication, and a commitment to education and continuous improvement.

3.2.2 Sustainable Finance Framework

Schoenmaker and Schramade (2018) presented a new framework for Sustainable finance based on the evolution of sustainable finance thinking which has undergone distinct phases in recent decades (refer to Table 3.1). The focus is gradually shifting from short-term profit to emphasizing long-term value creation. This book analyzes these stages and introduces a new framework for sustainable finance. Both financial and non-financial firms traditionally adhere to the shareholder model, where profit maximization is the primary goal. In Sustainable Finance 1.0 (as outlined in Table 3.1), financial institutions take the initial step of avoiding investments in companies with severe negative impacts on health (e.g., tobacco), international relations (e.g., cluster bombs), or the environment and wildlife/natural world (e.g., whale hunting). Some firms are now integrating social and environmental considerations into the stakeholder model (Sustainable Finance 2.0), highlighting the tension between the shareholder and stakeholder models. Questions arise about whether policymakers should permit a shareholder-oriented firm to acquire a stakeholder-oriented firm or if protection is needed for firms more advanced in sustainability. Another notable development is the shift from risk to opportunity. While financial firms initially avoided unsustainable companies from a risk perspective (Sustainable Finance 1.0 and 2.0), pioneers are increasingly investing in sustainable companies and projects to generate value for the wider community (Sustainable Finance 3.0).

The framework represents an important instrument for future understanding of sustainable finance and what should family business owners incorporate in their strategic approach to their organization in the future.

Table 3.1 Framework for sustainable finance

Sustainable finance typology	Value created	Ranking of factors	Horizon
Sustainable finance 1.0	Shareholder value	F > S and E	Short term
Sustainable finance 2.0	Stakeholder value	T = F + S + E	Medium term
Sustainable finance 3.0	Common good value	S and E > F	Long term

Note: F = financial value; S = social impact; E = environmental impact; T = total value
At Sustainable Finance 1.0, the maximization of F is subject to minor S and E constraints.
Source: Schoenmaker (2017)

3.2.3 Sustainable Finance Guidelines

Schoenmaker and Schramade (2018) also proposed a sustainable finance guideline for more successful implementation of the new Sustainability 3.0 in the organizations. Based on this guideline, we will propose the guidelines adopted for sustainable family businesses, as follows (Schoenmaker & Schramade, 2018):

Family Business Perspective

- Transition from a strictly profit-driven approach to embracing a stakeholder value model, ensuring a harmonious balance among the interests of family members, employees, suppliers, customers, and the community.
- Advocate for sustainable practices that contribute to long-term value creation, fostering shared benefits for the broader family and community.

Extended Family Involvement and Long-Term Vision

- Align the involvement of family members and their perspectives with long-term goals to counter short-termism.
- Implement reporting structures and compensation strategies that prioritize long-term success, including deferred rewards and mechanisms to recover incentives.
- Encourage transparent reporting on social and environmental initiatives to enhance accountability and align with family values.

Family Engagement

- Encourage family members to form coalitions and collaborate on sustainable initiatives, leveraging the collective influence of the family in addressing social and environmental challenges.

Sustainability Integration and Liquidity

- Promote awareness of alternative theories of sustainability within family businesses.
- Challenge the conventional perspective on liquidity, moving beyond immediate financial gains and considering the adaptive approach that adapts to changing environmental and social landscapes.

Governance and Family Values

- Minimize governance biases that favor short-term gains, allowing for long-term investments aligned with family values.
- Conduct stress tests to assess the family business's resilience to environmental challenges, emphasizing a proactive approach to sustainability with a focus on the family's long-term well-being.

These guidelines are tailored to empower family businesses to integrate sustainable practices, ensuring the longevity of the business and aligning with the values of the family.

3.3 Sustainable Finance Categories

3.3.1 Environmental (Green) Finance

In the realm of ESG, green finance allocates capital to innovative enterprises positively addressing climate issues or divesting from those contributing negatively. Negative green finance shifts investments from high to low carbon intensity or supports companies in reducing their carbon footprint, considering the long-term risks in petrochemical investments. Positive green finance invests in green technologies and environmentally sustainable practices, dominated by debt products like green bonds, which come in various forms such as corporate bonds, project bonds, asset-backed securities, and others.[1]

(i) Corporate bonds issued by a corporate entity to finance asset acquisitions.
(ii) Project bonds backed by single or multiple projects for which the investor has direct exposure to the risk of the project.
(iii) Asset-backed securities collateralized by one or more specific projects, usually providing recourse only to the assets.
(iv) Supranational, sovereign, sub-sovereign, or agency bonds issued by international financial institutions such as the World Bank or the European Investment Bank (EIB).
(v) Municipal bonds are issued by a municipal government, region, or city, which also includes sovereign bonds.
(vi) Finance sector bonds issued by a financial institution to raise capital to finance on-balance-sheet lending (such as loans) to green activities.

One of the widely embraced tools for advancing sustainable finance is green bonds—or corporate bonds issued by companies to fund or refinance their operations. These bonds, however, come with additional criteria specifically tailored to meet defined sustainability and environmental performance standards (Zamfiroiu, & Pînzaru, 2021). Green bonds play a pivotal role in channeling debt capital to support startups and high-growth ventures in green energy, water management, meat alternatives, and carbon capture. Additionally, there is a rising interest in blue bonds and blended finance models that focus on the "ocean economy," emphasizing biodiversity and marine sustainability. In 2020, sustainable debt issuance, including green bonds, totaled $732 billion, marking a 23% increase from the previous year.

3.3.2 Social (Impact Investment) Finance

Second, social finance provides start-up or growth capital into innovative enterprises that address a social market failure in the provision of welfare in sectors such

[1] Green bonds PP [f3] [lr].pdf (oecd.org).

as health, education, and employment (positive/integrated) or divests from companies that increase inequality of perpetuate social welfare failures (negative/exclusionary). Social impact bonds, integral to the realm of social finance, represent a novel form of payment by results agreement. Centered on achieving specific outcomes, these bonds involve collaborative efforts between public entities and private investors to fund initiatives addressing social issues. Social service providers, frequently in the form of social enterprise organizations, are commonly entrusted with delivering these services. The duration of such endeavors typically spans 3–5 years, although occasionally extending to longer periods (La Torre et al., 2019). Social bonds are also crafted with the explicit purpose of tackling or alleviating a specific social or environmental concern, often centered around a particular target demographic. In 2020, the International Capital Market Association (ICMA) introduced a comprehensive framework known as the Social Bond Principles. These principles encompass four fundamental components tailored to align with the articulated social or environmental objectives of the bond. These components encompass the financial utilization, the protocols for project assessment, financial management practices, and the reporting of impact (Nicholls, 2021). Social impact finance, challenges the traditional notion of bonds by serving as a contractually rooted collaboration focused on achieving specific outcomes. Unlike traditional fixed-rate securities, it acts as a funding mechanism contingent on predefined performance goals, embodying a contractually bound partnership among diverse stakeholders. These financial instruments integrate pay-for-performance and investment-driven strategies, involving private investors who provide upfront capital for interventions.

3.3.3 Governance (Stakeholder) Finance

The government plays a crucial role in fostering sustainable finance, but challenges persist in ensuring favorable returns in this sector. Key public policy changes necessary for advancing sustainable finance include revising legislation, incentivizing tax structures, establishing frameworks for capacity building, creating an environment for community investments, and allocating funds for sustainable goals in municipal plans. The adoption of these changes varies among governments, with underdeveloped countries facing challenges related to budget constraints and expertise (Rexhepi, 2016). Governance finance, linked to environmental or social finance, focuses on stakeholder investment, directing funds to companies following international standards for employee well-being. It involves investing in firms committed to integrating purpose-driven elements into their governance structures and assessing the impact on key stakeholders. Cooperatives and mutual organizations are pivotal in various impact sectors, including housing, agriculture, health, work integration, insurance, and banking. Global market share for mutual and cooperative insurers was 26.7% in 2017, with combined assets at $8.9 trillion (Nicholls, 2021).

3.4 Challenges Facing Sustainable Finance in Family Businesses

A fundamental goal of sustainable development involves the reduction of pollutants released into the atmosphere and the broader environment (Mirahsani et al., 2023). Small organizations, often faced with challenges, may prioritize economic benefits. Shepherd et al. (2013) found in their study that, despite positive attitudes toward environmental impact, entrepreneurs sometimes set aside these values to pursue opportunities that they know could harm the natural environment. Thus, having a sustainable business means being prepared for potential threats of all kinds (Mikušová et al., 2020). Research proved that opportunities and crises often have more impact on family businesses, those they need to be more aware of new risks and challenges (Abeysekera & Tran, 2021). Although family businesses have certain advantages they are confronted with specific problems, especially regarding governance, which in family businesses implies four crucial elements: family ownership and control, family management, multiple families and managers, and family succession, which usually could affect the family firm orientation (Hategan et al., 2019).

The challenges facing the integration of sustainable finance in family businesses extend beyond mere financial considerations and encompass the broader spectrum of organizational dynamics mainly influenced by regulatory frameworks, consumer expectations, and the need for environmentally responsible business practices. Addressing these challenges requires a holistic approach that considers management perspectives, employee engagement, and strategic initiatives to acquire, assimilate, and transform the family business's practices toward sustainability (Rexhepi et al., 2023). Furthermore, debates center around the control of sustainability finance initiatives, questioning whether the founding family or related family members should steer the financial direction toward sustainability (Fahlenbrach, 2009). Within the realm of sustainable finance in family businesses, the presence of a dedicated team of environmentally conscious employees assumes a crucial role. Many of these individuals have been integral to the company since its inception, growing and evolving in tandem with its green finance initiatives. Sustainability finance practices within the family business are often fortified through familial bonds and traditions. Family members who share the vision of green finance objectives frequently occupy influential positions in the business, showcasing a heightened ability to adapt to an environmentally conscious workplace culture compared to their unrelated colleagues. They, often have direct access to key decision-makers, fostering a collaborative approach, while other employees adhere to formal channels (Jahmurataj et al., 2023).

Family businesses in order to implement concepts that could improve the environmental impact need to have a proper strategy for that. Given that the success of an organization significantly hinges on how its employees think and behave (a crucial aspect of meeting sustainable development objectives lies in fostering environmentally conscious behaviors and practices among employees (Mirahsani et al., 2023). Numerous empirical studies affirm that various expressions of Employee

Green Behavior (EGB), including organizational citizenship behavior toward the environment and the development of environmentally friendly products, contribute positively to the environmental performance of organizations. The voluntary adoption of green behaviors by employees within a family business, such as actively sharing knowledge on environmental protection, identifying ecological issues, and providing suggestions to management, can significantly enhance the organization's environmental performance.

As family businesses strive to integrate sustainable finance into their operations, creating a workplace culture that encourages and supports environmentally responsible practices becomes vital. This extends beyond organizational-level policies to encompass individual behaviors and attitudes, reinforcing the notion that sustainable development is not only a strategic objective but also a collective responsibility shared by all members of the organization, including its employees (Mirahsani et al., 2023). Notably, the level of effort invested in creating conducive workplace conditions for employees to actively participate in voluntary green behaviors stands out as a crucial challenge in realizing sustainability goals within the family business. This underscores the importance of fostering a green culture within the organization, aligning individual behaviors with environmental objectives, and contributing collectively to the family business's commitment to sustainability through the integration of green finance strategies (Mirahsani et al., 2023). In a study involving 482 SMEs in Pakistan, conducted interviews with 384 owners/managers, analyzing the data through Structural Equation Modeling. The findings concluded that Intellectual Capital significantly impacts the Sustainability Performance of companies. Specifically, the study indicated that more experienced managers correlate with more pronounced corporate sustainability, emphasizing the pivotal role of human capital in the sustainable practices of businesses, including family enterprises (Kajtazi et al., 2023). It is obvious that establishing a green organizational culture in a family business will result in a more environmentally friendly and resource-efficient approach, aiming for reduced environmental pollution and increased utilization of renewable sources. There are other challenges related to sustainable finance such are (Nicholls, 2021):

1. **Market Efficiency**—An important challenge related to sustainability finance is related to the supply side or the lack of investable deals in the sector. Even though there might be opportunities in green and sustainable finance, the demand side of deals related to SDGs is not clearly defined. Also, there are currently many high transaction costs in putting together many deals and funds because of the lack of a robust impact performance evidence base and data, as well as the complexity of blended transactions. All markets function best on abundant and robust performance data.
2. **Standards and Data**—A primary challenge is the absence of standardized metrics and reporting frameworks, impeding consistent ESG measurement and comparability (Scholtens, 2020). Currently, there is a lack of data and standards of investment related to sustainability, and there are no agreed standards on regulations related to sustainable financing, even though recently there are many initia-

tives that are addressing these issues. In 2019, in collaboration with the Impact Management Project, under the UNDP, developed a set of standards for investing under the SDG Impact initiative.

3. **Monetization** remains a key challenge since almost all organizations tend to maximize their profitability, this is especially the case of family businesses that are small and cannot invest in sustainability-oriented projects which are perceived as additional expenses to these organizations. Numerous investigations, spanning various domains, indicate the existence of opportunities for ESG outperformance (Friede et al., 2015). Companies excelling in ESG criteria are not only champions in sustainability but also entities with effective strategies and practices (Zamfiroiu & Pînzaru, 2021). Thus, showing more cases of positive impact from the implementation of sustainability programs might further increase the use of sustainable finances by incorporating more sustainable innovations.
4. **Green washing is perhaps the most faced challenge.** Sustainable development is a contend that addresses trade-offs between the ambition of economic, social, and environmental goals, and the feasibility of achieving them requires acknowledgment of biophysical, social, and political constraints which is related to the business owner's ethical standards (Zamfiroiu, & Pînzaru, 2021). Concerns regarding "greenwashing" have surfaced, where entities exaggerate or misrepresent their sustainability efforts. This underscores the need for enhanced regulation and enforcement to ensure transparency and authenticity (Bauer & Hann, 2018). According to the World Economic Forum while 63% of the companies in its ESG index had a policy in place to reduce their emissions, only 35% had specific reduction targets. Finding ways to control the missuses of green elements into more increase of the profitability of the organizations will remain a key challenge to solve.
5. **Family Business Values**—Unlike human capital, which is embodied in individuals, social capital is embodied in relationships among people and formal social institutions (Danes et al., 2008). Sustainable finance is related to family's beliefs about what is a value and where do they see their organizations. Thus, family businesses may hold distinct perspectives of socially responsible business behavior (Astrachan et al., 2020; Niehm et al., 2008). Family business values are defined as clear and desirable goals for both family and business lives. Given that there are often conflicts of interest between business and family goals, family business values should be well known and serve as a common ground between the business and the family to achieve their goals (Oudah et al., 2018).

Considering all of these challenges and incorporating them into family business strategy may have a serious impact on the successful implementation of sustainable finance. The renowned Harvard Business School professors Michael Porter and Mark Kramer, in their influential article "The Big Idea: Creating Shared Value" (2011), assert that corporate strategies must align with societal needs. Contrary to traditional thinking, pursuing social or environmental objectives need not necessarily involve a financial trade-off. Experiences in the market have demonstrated that, in certain areas, social investments can yield both solid financial returns and social impact. In these domains, social investors emerge as crucial contributors, providing private capital to address social challenges in innovative ways (Wilson, 2014). In the United States, as of the end of 2020, 33% of assets under management were reportedly directed toward ESG strategies. These strategies, distinct from an agnostic stance, involve approaches that explicitly consider environmental, social, or governance factors in their decision-making (Chiu et al., 2022).

3.5 Integration of Sustainable Finance Reporting in Family Businesses

The successful implementation of any novel approach relies on effective control or, in our context, reporting. Major institutions associated with ESG recommend the integration of sustainability reports into a broader array of organizations. In the European Union, this is poised to become mandatory for the majority of organizations. Another pivotal principle emphasizes the promotion of transparency and accountability (Gibson & Pivo, 2019). Whether they are companies or financial institutions, entities are encouraged to disclose details pertaining to their ESG (Environmental, Social, and Governance) performance. This revelation allows stakeholders to assess their commitment to sustainable finance objectives. Such transparency empowers investors and consumers, enabling them to make informed decisions and demand accountability from organizations concerning their sustainability commitments (Bauer & Hann, 2018).

The origins of green finance reporting can be traced back to the evolution of financial reporting practices, which encompassed social reports in the 1960s, environmental reports in the early 2000s, and non-financial reporting in the 1980s (Sepasi et al., 2018). The amalgamation of social and environmental disclosures gave rise to the first quasi-sustainability reports in the 1990s (Moss et al., 2015). Notably, the push for standardization in green finance reporting gained traction with the establishment of the Green Finance Reporting Initiative. GFRI introduced its inaugural reporting framework in 2018, with subsequent revisions in 2020 and 2022. The most recent versions, GFRI G4 and GFRI Standards, were unveiled in 2024 and 2026, respectively, solidifying GFRI's position as a leading framework for the voluntary reporting of green finance information on a global scale (Sepasi et al., 2018). Recent studies reveal an increasing trend where financial organizations, after the financial crisis, exert pressure on corporate executives to furnish financial reports encompassing non-financial performance. This underscores the pivotal role of social finance in striking a balance between organizational profitability, keeping at a moderate level, and fostering employee motivation.

Sustainability reporting incorporates three elements which also have some sub-elements:[2]

[2] https://www.elizabethjoss.com/the-elements-of-sustainability-reporting/.

1. Governance
   - Approach to sustainability.
   - Internal assessments.
   - External research.
   - Materiality assessment highlighting stakeholders' concerns about sustainability and ESG topics.
   - Stance on ESG matters and sustainability strategy, as well as ambitions and goals.
   - Sustainability roadmap, compliance, and measurement efforts, along with external sustainability initiatives and how they align with internal regulations and frameworks.
   - Internal controls, policies, and procedures governing leadership, board composition, sustainability committee responsibility, and whistleblower programs.
   - Shareholder rights, audits, committee structures, efforts to combat corruption, and sustainable investments for the future.
2. Environment
   - Impact on the environment and vice versa.
   - Efforts toward carbon neutrality, water reduction, waste management, and other environmental aspects.
   - Dedication to address environmental topics such as:
     - Climate change.
     - Carbon emissions.
     - Biodiversity.
     - Air pollution.
     - Sustainable resources in the supply chain.
3. Social
   - Fair and equal wages.
   - Employee engagement.
   - Creating an open and transparent work environment.
   - Diversity, equity, and inclusion (DEI) efforts.
   - Awards your company has received on social-related topics.
   - Community involvement or outreach projects.
   - Labor standards, human rights, data protection, privacy, and gender-related initiatives.

Many organizations have embraced the integration of sustainable finance into their organizational structures, incorporating it across various aspects such as financial strategies, operations, stakeholder engagement, and reporting. Sustainability finance reporting emerges as a robust mechanism for family businesses to evaluate their financial performance, convey their sustainable financial practices to stakeholders, and contribute to broader societal awareness.

GFRI's primary objective is to promote a standardized approach to green finance disclosures by establishing a universally accepted framework for reporting. As of the latest update, approximately 5000 family businesses have issued around 15,000 reports utilizing GFRI guidelines. Following the widespread adoption of GFRI green finance reporting, an increasing number of family businesses have initiated the reporting of their sustainable financial practices (Smith et al., 2018).

3.6 Summary

The current challenges in the business environment have spurred significant interest in areas such as green and sustainable finance, corporate social responsibility, and both financial and non-financial performances. These factors are now pivotal in gauging the success of organizations, determining their competitive advantage, and assessing their impact on the marketplace. Notably, the interplay of green and sustainable finance, corporate social responsibility, and intellectual and human capital has assumed a central role in shaping organizational outcomes and influence in the contemporary business landscape (Popescu & Popescu, 2019). Sustainable finance in family businesses is more than just a trend; it is a fundamental shift toward responsible stewardship of both financial and environmental resources. Family businesses that embrace this approach can look forward to a future in which they not only prosper financially but also leave a positive, lasting impact on the world. As they navigate the complex landscape of sustainable finance, family businesses should remain committed to their values, adapt to changing circumstances, and continue their journey toward intergenerational equity.

Questions for Discussion

- How can family businesses effectively integrate sustainable finance principles into their overall financial strategy and decision-making processes?
- What are the key success factors for family businesses in implementing sustainable finance practices, and how do these practices contribute to the long-term viability and resilience of the business?

Case Study: ING Provides Sustainable Loan to Philips

In April 2017, Philips, the healthcare technology company, entered into an innovative €1 billion loan agreement with a consortium of banks. This unique facility features an interest rate tied to Philips' year-on-year sustainability performance. The structure of the revolving credit facility means that if Philips enhances its sustainability performance, the associated interest rate decreases, and conversely, if performance declines, the interest rate increases. The 5-year loan, set to mature in April 2022, is designated for general corporate purposes.

Distinguishing itself from prior green loans and bonds, where pricing was linked to specific green covenants or funds were restricted to green initiatives, this loan offers a dynamic connection to sustainability performance.

Among the 16 banks in the consortium providing the loan, ING Bank assumes the role of conducting the credit risk assessment and serving as the sustainability coordinator for the syndicated loan. Sustainalytics, an independent provider of environmental, social, and corporate governance research and ratings, has evaluated and benchmarked Philips' current sustainability performance.

ING highlights that the loan agreement with Philips serves as an additional means for the bank to support and reward clients striving for greater sustainability. This loan initiative aligns with Philips' "Healthy People, Sustainable Planet" program, wherein the company aims to achieve carbon neutrality in its global operations and source all electricity from renewable sources by 2020 (SDG 12). Additionally, Philips seeks to enhance the lives of 3 billion people annually by 2025 through innovative solutions that contribute to a healthier and more sustainable world (SDG 3).

Questions:

1. How do you see the decision of ING Bank to finance Philips in terms of sustainability?
2. What kind of other sustainable financing sources do you propose to Philips?

Source: Adopted from INGand Philips collaborate on sustainable loan I ING.

- What challenges and barriers do family businesses commonly face when adopting sustainable finance, and how can these challenges be overcome or mitigated?
- How do family businesses engage with various stakeholders, including family members, employees, and the broader community, to ensure alignment and support for sustainable finance initiatives?
- What metrics and indicators are most relevant for family businesses to measure the impact of sustainable finance initiatives, and how can they effectively communicate these impacts to both internal and external stakeholders?

Additional Readings

- Schoenmaker, D., & Schramade, W. (2018). *Principles of sustainable finance*. Oxford University Press.
- Lehner, O. M. (Ed.). (2016). *Routledge handbook of social and sustainable finance*. Routledge.
- Rexhepi, G. (2016). The architecture of social finance. *Routledge handbook of social and sustainable finance*, 35–49.

Suggested Activities

- In small groups, students should develop a comprehensive sustainability plan for a hypothetical family business. The plan should address environmental, social, and governance aspects, outlining specific initiatives, targets, and key performance indicators. Groups should present their plans to the class, emphasizing the feasibility and impact of their proposed strategies.

- Assign students to small groups and task each group with developing a sustainable finance strategy for a hypothetical family business.
- Encourage students to discuss the negative impact and positive impact of implementation of sustainability in family businesses.
- Select a real-life family business that has faced sustainability challenges or successfully implemented sustainable practices and discuss them in small groups.

Keywords and Concepts

- Sustainable finance
- Family businesses
- Green finance
- Social finance
- Environmental dimension
- Economic dimension
- Governance (ESG) criteria
- Sustainable Development Goals (SDGs)
- Triple-bottom line
- Resource conservation
- Philanthropy
- Global poverty
- Financial redistribution
- Environmental sustainability
- Principles of sustainable finance

References

Abeysekera, I., & Tran, K. T. (2021). The coronavirus as a disrupter of a sustainable small early childhood family business in Vietnam. *Sustainability, 13*(19), 10692.

Astrachan, J. H., Binz Astrachan, C., Campopiano, G., & Baù, M. (2020). Values, spirituality and religion: Family business and the roots of sustainable ethical behavior. *Journal of Business Ethics, 163*, 637–645.

Basak, M. (2019). Integrating ESG in family businesses: The strategic role of family values. *International Journal of Family Business, 14*(3), 187–204.

Bauer, R., & Hann, D. (2018). Corporate environmental management and credit risk. *Journal of Business Finance & Accounting, 45*(9–10), 1088–1115.

Brown, A., & Miller, J. (2017). Sustainable finance and family businesses: A holistic approach. *Journal of Sustainable Finance & Investment, 7*(2), 154–170.

Chiu, I. H., Lin, L., & Rouch, D. (2022). Law and regulation for sustainable finance. *European Business Organization Law Review, 23*(1), 1–7.

Danes, S. M., Lee, J., Stafford, K., & Heck, R. K. Z. (2008). The effects of ethnicity, families and culture on entrepreneurial experience: An extension of sustainable family business theory. *Journal of Developmental Entrepreneurship, 13*(03), 229–268.

Davis, P. J., et al. (2020). *Sustainable finance and investments: Implementing ESG strategies for long-term value*. Wiley.

European Commission. (2015). Green paper – Building a capital markets union.

Fahlenbrach, R. (2009). Founder-CEOs, investment decisions, and stock market performance. *Journal of Financial and Quantitative Analysis, 44*(2), 439–466.

Friede, G., Busch, T., & Bassen, A. (2015). ESG and financial performance: Aggregated evidence from more than 2000 empirical studies. *Journal of Sustainable Finance & Investment, 5*(4), 210–233.

Gibson, K., & Pivo, G. (2019). The valuation implications of sustainability in real estate markets. *Journal of Real Estate Literature, 27*(3), 279–308.

Gomez, L., & Anderson, M. (2018a). Sustainable finance initiatives in family businesses: A path to long-term viability. *Journal of Family Business Management, 8*(2), 214–230.

Gomez, R., & Anderson, S. (2018b). *Responsible ownership: Resolving the agency problem in family firms*. Palgrave Macmillan.

Hategan, C. D., Curea-Pitorac, R. I., & Hategan, V. P. (2019). The Romanian family businesses philosophy for performance and sustainability. *Sustainability, 11*(6), 1715.

Jahmurataj, V., Ramadani, V., Bexheti, A., Rexhepi, G., Abazi-Alili, H., & Krasniqi, B. A. (2023). Unveiling the determining factors of family business longevity: Evidence from Kosovo. *Journal of Business Research, 159*, 113745.

Kajtazi, K., Rexhepi, G., Sharif, A., & Ozturk, I. (2023). Business model innovation and its impact on corporate sustainability. *Journal of Business Research, 166*, 114082.

Kramer, M. R., & Porter, M. (2011). *Creating shared value* (Vol. 17). FSG.

La Torre, M., Trotta, A., Chiappini, H., & Rizzello, A. (2019). Business models for sustainable finance: The case study of social impact bonds. *Sustainability, 11*(7), 1887.

Lund, D., & Myers, S. (2020). The role of green finance in family business sustainability. *Family Business Review, 33*(2), 179–197.

Mikušová, M., Friedrich, V., & Horváthová, P. (2020). Who is more sustainable? Family business or non-family business? *Czech evidence. Sustainability, 12*(14), 5540.

Mirahsani, N., Azizan, O., Shahriari, M., Rexhepi, G., & Najmi, A. (2023). Green culture toward employee green behavior; the mediation roles of perceived support and green identity. *Environment, Development and Sustainability, 30*(1), 1–24.

Moss, T. W., Neubaum, D. O., & Meyskens, M. (2015). The effect of virtuous and entrepreneurial orientations on microfinance lending and repayment: A signaling theory perspective. *Entrepreneurship Theory and Practice, 39*(1), 27–52.

Murray, A., & Williard, M. (2018). *The ESG factors in sustainable investing: From principles to practice*. Springer.

Nicholls, A. (2021). *Sustainable finance: A primer and recent developments* (pp. 1–51). Asian Development Outlook.

Niehm, L. S., Swinney, J., & Miller, N. J. (2008). Community social responsibility and its consequences for family business performance. *Journal of Small Business Management, 46*(3), 331–350.

O'Donohoe, N., Leijonhufvud, C., Saltuk, Y., Bugg-Levine, A., & Brandenburg, M. (2010). Impact investments. *An Emerging Asset Class, 96*, 1–96.

Oudah, M., Jabeen, F., & Dixon, C. (2018). Determinants linked to family business sustainability in the UAE: An AHP approach. *Sustainability, 10*(1), 246.

Popescu, C. R. G., & Popescu, G. N. (2019). An exploratory study based on a questionnaire concerning green and sustainable finance, corporate social responsibility, and performance: Evidence from the Romanian business environment. *Journal of Risk and Financial Management, 12*(4), 162.

Rahdari, A., Sheehy, B., Khan, H. Z., Braendle, U., Rexhepi, G., & Sepasi, S. (2020). Exploring global retailers' corporate social responsibility performance. *Heliyon, 6*(8), e04644.

Rexhepi, G. (2016). The architecture of social finance. In *Routledge handbook of social and sustainable finance* (pp. 35–49). Routledge.

Rexhepi, G., Abazi-Alili, H., Abduli, S., Ibraimi, S., & Zuferi, R. (2023). Green entrepreneurship and firm performance: The case of Albania. In *Entrepreneurship development in the Balkans: Perspective from diverse contexts* (pp. 69–80). Emerald Publishing Limited.

Rexhepi, G., Kurtishi, S., & Bexheti, G. (2013). Corporate social responsibility (CSR) and innovation–the drivers of business growth? *Procedia-Social and Behavioral Sciences, 75*, 532–541.

Rexhepi, G., Ramadani, V., & Ratten, V. (2018). TQM techniques as an innovative approach in sport organisations management: Toward a conceptual framework. *International Journal of Business and Globalisation, 20*(1), 18–30.

Schoenmaker, D. (2017). *Investing for the common good: A sustainable finance framework* (p. 80). Bruegel.

Schoenmaker, D., & Schramade, W. (2018). *Principles of sustainable finance*. Oxford University Press.

Scholtens, B. (2020). The future of sustainable finance. *Review of Financial Economics, 39*(1), 14–18. European Commission.

Sepasi, S., Rahdari, A., & Rexhepi, G. (2018). Developing a sustainability reporting assessment tool for higher education institutions: The University of California. *Sustainable Development, 26*(6), 672–682.

Shepherd, D. A., Patzelt, H., & Baron, R. A. (2013). "I care about nature, but…": Disengaging values in assessing opportunities that cause harm. *Academy of Management Journal, 56*(5), 1251–1273.

Wilson, J. (2014). *Essentials of business research: A guide to doing your research project* (pp. 1–376). Essentials of Business Research.

Zamfiroiu, T. P., & Pînzaru, F. (2021). Advancing strategic management through sustainable finance. *Management Dynamics in the Knowledge Economy, 9*(2), 279–291.

[illegible]
[illegible] (2018) [illegible]
[illegible]
[illegible]

Sustainable Human Resource Management in Family Businesses

4

Learning Objectives
After reading this chapter, you should be able to:

- Define sustainable human resource management in family business.
- Analyze sustainable HRM settings in family businesses.
- Address the importance of sustainable HRM in family business.
- Understand tools and mechanisms of sustainable HRM in family business.

Profile: LEGO
LEGO was founded by Ole Kirk Kristiansen in 1932, and it is now one of the world's largest toy-producing companies. LEGO is the ideal instance where a multi-million company uses the triple-bottom-line approach to shape its sustainable Human Resource Management.

In reference to the "people" objective of the triple-bottom line, LEGO's Human Resource Management ensures that their workplaces are inclusive and diverse, and only ethical business behavior is present between the employees and customers. Diversity and inclusivity incorporate workers of different ages, genders, education, and ethnic or cultural backgrounds. LEGO believes that improving its HR system and processes is crucial to triumphant recruitment and career development. During the year 2020, LEGO presented extended parental leave for their employees worldwide. Since then, all employees, globally, can get at least 26 weeks of paid parental leave if they are considering themselves primary caregivers. Otherwise, secondary caregivers are allowed to receive 8 weeks of paid leave. In addition to these capabilities, the company guarantees monetary

R. Palalić et al., *Sustainable Family Business*, Springer Texts in Business and Economics, https://doi.org/10.1007/978-3-031-57999-8_4

security for their employees in the case of deaths in the family or permanent disability.

"Our ambition is to make LEGO® bricks from more sustainable sources without compromising on quality or safety."

LEGO has taken several steps toward making their organization environment friendly, including getting into partnership with the WWF (World Wildlife Fund), reducing their carbon dioxide emissions, and producing sustainable packaging. Sustainable packaging refers to producing LEGO boxes from cardboard, with 75% of it coming from recycled materials and can be recycled in infrastructures where access to cardboard recycling is available. As LEGO invested in producing renewable energy from the offshore winds in Germany and the United Kingdom, the organization now uses 100% renewable energy for producing their toys. Currently, more renewable energy is being produced than used up from their workplaces all over the world.

In order to implement these strategies, LEGO took a huge financial risk. The first adaptation of their products to sustain the environment cost them a profit drop of approximately 18%. Fortunately for them, they did not get unmotivated by this and continued their environment-friendly production processes, and in 2020, they reached a profit margin of 13%. Compared to their profit before they adapted such policies, the profit in 2020 was much higher.

Using the tripe-bottom-line approach may be costly and risky for businesses, but LEGO has proved that profit is attainable even while using this strategy. The principal idea behind this strategy is to maximize a business's profits while ensuring a sustainable environment and protection of the employees and customers. Despite the time needed for the results of this strategy to show, it will have a great impact on the environment and both businesses and people will prosper.

Source: Authors.

Bibliography

1. https://www.linkedin.com/pulse/tbl-case-study-lego-aanandita-singhal, accessed on September 3, 2023
2. https://www.lego.com/cdn/cs/aboutus/assets/bltf291ca87cee388b5/HR_Diversity_and_Inclusion_policy.pdf, accessed on September 3, 2023
3. https://diginomica.com/how-lego-shapes-long-term-hr-thinking-brick-brick, accessed on September 3, 2023
4. https://www.lego.com/en-at/sustainability/environment, accessed on September 3, 2023
5. https://www.lego.com/en-at/sustainability/environment/sustainable-materials, accessed on September 3, 2023

4.1 Introduction

The most valuable asset that both family-owned and non-family businesses have is their workforce (human resources) (Ramadani & Hoy, 2015; Zellweger et al., 2010), Zellweger et al., 2012). In order to prevent human resources from becoming obsolete, these assets should be nurtured, treasured, and employed in a balanced way. In other words, in order to prolong their life over the long term, human resources must be "kept" in the same manner as other tangible assets, such as machinery, equipment, and so forth.

Many studies conducted over the past 10 years have shown how important sustainability is to human resource management. But the more strongly proclaimed the need for this sustainability, the more difficult it became. According to some research (Aust et al., 2020), if the company wants to have sustainable human resource management (HRM), it should change its paradigm toward strategic HRM, which paves the way for long-term sustainability. The triple-bottom-line approach to sustainability is the lens through which strategic HRM sustainability is being seen as the great facilitator (Bush, 2020). By providing visionary perspectives through which a company (family or non-family business) can see better *people*, *planet*, and *profit*, HRM, especially SHRM, enables businesses to realize the sustainability agenda (Podgorodnichenko et al., 2020) over the long term.

The purpose of this chapter is to explain the nature of human resource management, its methods and mechanisms in family businesses, and why HRM should be viewed as strategic human resource management (SHRM) in the long perspective. In addition, we present potential models that, in the context of the TBL approach, can improve the paths toward the sustainability of family companies.

4.2 Nature of HRM in Family Businesses

One of the main pillars in every organization is human resource management department. Because human capital is critical that can bear all challenges firms face with. Human capital or human resources are backbone of the firm and thus by its nature, it plays the most important role in the firm's development and growth. Moreover, a prodigy human capital possesses talents, skills, knowledge, and know-how, which are synergized in firm's creativity and innovation. Yet, the development of new products or services is emergent on the human capital or human resources.

As we discussed in Chap. 2, to be self-sustainable, each organization needs to achieve the competitive advantage in the long term. So do family businesses. The source of sustainability in family business lies in human resources in first place too, because people, the human part of the firm is its stamina without which any operational activity cannot be done.

4.3 Lopez-Cabrales and Valle-Cabrera's HRM Strategies for TBL Approach[1]

Strategic human resource management sustainability lies in visionary leadership that can see new business horizons (Lopez-Cabrales & Valle-Cabrera, 2020). In other words, the leadership in family business should be determined to seek the best that concerns the company (profit), society (internal and external stakeholders), and environment (planet). Accordingly, family business owners, or so-called "big fathers" (Ramadani et al., 2020) should be the ones who make the choice to pursue the firms' sustainability.

Undertaking a strategy that will implement the TBL approach is the true strategy that can help the business to be challenged and to bring the challenges to other competitors. Being in line with the TBL approach, family businesses are becoming dynamic, responsive, adaptable, and more proactive, which gives them a better chance to gain and maintain a competitive advantage over their competitors.

Strategic human resource management can help family business to achieve a competitive advantage using the TBL approach. In the first place, as the HRM department, it deals with overall human capital (human resource) development. Therefore, it requires investment in human resources on a regular basis, depending on the nature of the business family business is in, as well as the current status of knowledge, skills, and know-how of people in the firm.

One of the ways that HRM can contribute and be in line with the TBL approach is the development of better **employment relationships** ensured by means of the HRM department. Competing in the market, there is no difference between non-family and family business. Therefore, the above model discussed in the next paragraphs could be very useful in creating a sustainable path toward long-term sustainability using TBL approach in family business, though.

The model that enhances the long-term contribution of HRM to triple-bottom-line approach are proposed by Lopez-Cabrales and Valle-Cabrera (2020). According to Lopez-Cabrales and Valle-Cabrera (2020), the models propose the role of HRM which can be integrated into sustainable corporate strategy through strategic HRM. These models of employment relationships are *two balanced* models that include:

1) Quasi-Spot Contract
2) Mutual Investment

Additional two are *imbalanced* ones:

3) Over-investment
4) Underinvestment (p. 6)

[1] This subsection is based on a theory proposed by Lopez-Cabrales and Valle-Cabrera (2020).

Table 4.1 Employment relationship models

Employment relationship		
Approach	Models	Implications
Imbalanced 1	Over-investment	A high level of investments in human component but lower-level expectations from employees. Goes beyond the ordinary level of investment that paves the way for long-term sustainability based on creativity, innovation, proactivity, etc. that leads to long-term triple sustainability.
Imbalanced 2	Underinvestment	Low level of investment in human resources, but high expectations from employees in return.
Balanced 1	Quasi-spot contract	Short-termed, fair approach, with balanced incentives and expectations.
Balanced 2	Mutual investment	Mutual agreement on a high-level investment in human component with the same level of expectations from employees.

Each of the models is important for family businesses to oversee the current stage of the HRM development in terms of employment relationships. Table 4.1 represents the models briefly.

4.3.1 Balanced Approach 1: Quasi-Spot Contract Employment Relationship

We know that any procedure, action, regulation, or rule in the company is aimed at shaping employees' behavior so that can enhance the overall performance of the company. This model is intended to offer a limited set of incentives and expects a minimal level of behaviors associated with sustainability in exchange. This is a short-term contract where both parties, the firm and employee, are agreed upon. Such rules being imposed, or the contract is perceived as fair by employees (Tsui & Wu, 2005) because it provides a minimum of stimuli and asks for the minimum in return. This shows compliance with the HRM strategies too.

4.3.2 Balanced Approach 2: Mutual Investment Employment Relationship

Lopez-Cabrales and Valle-Cabrera (2020) argue that this approach is opposite to the Quasi-Spot Contract one. Actually, here the firm offers a high level of stimuli or incentives while expecting the same level of contributions and commitments from employees. The stimuli' range covers economic, social, and environmental (TBL) perspectives. For example, economic incentives can be referred to salaries in the first place, which are economically and socially competitive in the market. Social perspective in this case can be seen in providing professional development at a higher level and introducing the work-life balance to employees while environmental considerations can be implemented through high-level workshops and seminars on employees' awareness of environmental issues so that employees could implement this knowledge in their daily tasks. More precisely, employees can be taught to try as much as possible to use environmentally friendly resources in their activities. This model or approach can be called "Mutual-Sustainability" (Lopez-Cabrales & Valle-Cabrera, 2020) as well.

4.3.3 Imbalanced Approach 1: Under-Sustainability Investment Employment Relationship

On the contrary of the previous model(s), this model is made of imbalance between the investment provided in human resources and expectations. It denotes that a very low investment has been made but expectations are high in terms of commitment and performance of employees. Firms that implement this model, they actually do not see their long-term sustainability as imperative. They concentrate on one component (profit), while neglecting two others. However, the question is whether the firm is able to perform well by applying poor HRM practices.

Similarly, such a firm does not have a competitive advantage because it is short-sighted while not considering the development of the proper HRM system that will lead to a competitive advantage. As the incentives are low, and expectations are high, there is a high likelihood of a high-rate employee turnover, which makes the firm's competitiveness weaker and underdeveloped. Additionally, in this model the HRM practices are minimized or no HRM practices, then all three components are underdeveloped and do not represent. Employees do not observe or give any economic, social, or environmental perspective. On the opposite, the HRM should be designed and be effective in providing long-term sustainability (Aust et al., 2020), because it is a global issue and yet competitiveness is imperative to all types of firms. This model can be also called Under-Sustainability.

4.3.4 Imbalanced Approach 2: Over-Sustainability Employment Relationship

This model can be called "Over-investment" (Lopez-Cabrales & Valle-Cabrera, 2020; Tsui & Wu, 2005). The premise of this model lies in the fact that there is an imbalance between the investment and employee's responses. In other words, investment in human component outweighs the actual return from employees. Such a high investment in the firm is not only related to economic perspective of the employment relationship, but investment related to sustainable future competitive advantage in terms of "innovation, proactiveness, prevention" etc. (Lopez-Cabrales & Valle-Cabrera, 2020), and competitiveness depends on. Special attention is given to "Innovative Sustainable HRM System."

The other side of the coin of this over-investment model is that the family firm will attract high-quality employees which will keep the human resource pool wealthy and ready to take challenges in this TBL approach. Although the firms are aware of such an over-investment, this will help to implement this triple sustainability in the long term.

4.3.5 Implication of the Employment Relationship Model

The models are represented in Table 4.1 explained to us possible options that family businesses can pursue in order to implement and maintain the triple-bottom-line model. The following implications can be derived from these models. Figure 4.1 shows in a nutshell how the models go to sustainability achievement through the triple sustainability (TBL approach).

Implications are reflected in terms of types of HRM strategies that enhance the triple sustainability.

Implication one: In the case of *Underinvestment* model, the firm has a problem regarding the HRM policies that can help in developing sustainable HRM. Most likely that such a firm takes the posture "wait-and-see" (Barringer & Ireland, 2019), which denotes a traditional risk-averse family business (p. 33). The corporate strategy is weak with "inertia" (Rothaermel, 2017) that pertains to organizational inertia (p. 234). Moreover, there is a missing HRM strategy because there is no investment in the HRM strategy.

Implication two: *Quasi-Spot Contract* model shows a kind of fair play between the firm or HR system and employees. It is somehow a balanced HR system but not enough so that it can be regarded as long-term HR.

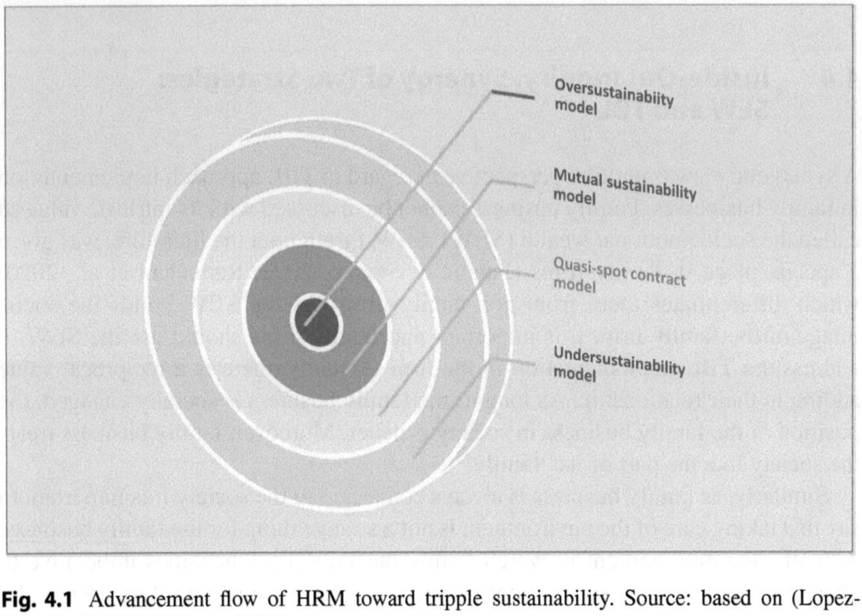

Fig. 4.1 Advancement flow of HRM toward tripple sustainability. Source: based on (Lopez-Cabrales & Valle-Cabrera, 2020)

Implication three: *Mutual Investment Sustainability* model shows an advanced HR system where the policies are made to ensure "mutual" commitment between the firm and employees. The (family) firm provides high incentives that cover economic, social, and environmental aspects while expecting a higher return from employees in regard to positive performance outcomes.

Implication four: *Over-sustainability model* represents an extra mile in the development of HR policies that provide additional investment beyond employee expectations. However, employees will not be able to replicate the same value to the firm as the return, but in the long-term will be reflected in the innovative products and services that will be sustainable covering all three sustainability components while maintaining and ensuring the family firms' sustainable competitive advantage.

Figure 4.1 summarizes the development of the HR system that will stream toward triple-bottom-line approach which will give family firms greater competitive competencies and advantages.

In Figure 4.1, we see that the heart for the long-term triple sustainability is actually a positively imbalanced approach, the over-sustainability for which family firms need to invest more than they can get in return from employees in the short term.

Sometimes, family firms are not able to invest more than they have, but they can offer for a while a mutual agreement on incentives and expected outcomes.

In the third circle, quasi-spot model is actually no pain-no gain model, but the HRM is not close to advancement due to internal firm's circumstances.

The last circle shows underdeveloped HRM where human capital component is suffering due to no investment in it. These family firms can be considered disoriented and have no intention toward triple-bottom-line strategy.

4.4 Inside-Out Inquiry: Synergy of Two Strategies: SEW and TBL

A synergetic view is another perspective in regard to TBL approach implementation in family businesses. Family business cannot be discussed with its intrinsic value so called the socioemotional wealth (SEW). SEW, throughout the literature, was given a special place in family firms (Palalić & Smajić, 2022; Ramadani et al., 2020) which differentiates them from non-family firms. As the SEW builds the social image of the family firm, it is important that family firms should use the SEW to address the TBL implementation in the firm. Actually, there is a reciprocal value adding in their relationship. As long as the family business is socially engaged, the position of the family business in society is better. Moreover, family business treats the society like the part of its "family."

Similarly, as family business is always connected to the society it is important to say that taking care of the environment is not a strange thing for the family business. Actually, the environment in which family business plays becomes imperative to cherish it and protect from different perspectives. Environmental safety ensured by the family firm is in fact additional block in a positive image of the family business in the society. Thus, family businesses should always strive to have these TBLs in the agenda when creating their strategic plan.

Finally, having all said above, the emotional part of the SEW provides a good tool to establish and devote resources to TBL implementation in family firms. Both play on emotions, the TBL (society, environment, and people), and the family firms. Playing on emotions means that having empathy toward the planet is crucial to helping everyone on it as well as earning some profits. Literally, the TBL is strong leverage to family business to create a fantastic strategic agenda that will fit all (the family firm and the society) while not ruining the long-term profit because the TBL, as the awareness of this phenomenon is increasing around the globe, will bring the value to the family business (the profit). Hence, the synergetic approach is worthwhile which leaves positive traces for the new generations.

4.5 Future Benefits of Implementation HRM's TBL

TBL being implemented in the HRM in a family firm implies opening another perspective. This is family firm branding. Family branding is an outstanding reputation of the family firm in the industry and society that signifies the importance of the family's role in socioeconomic development. The importance of a family firm's image to external stakeholders is paramount because family businesses are kin to establish a good family reputation in the society via "community embeddedness" (Lumpkin & Bacq, 2022). With implemented HRM policy, where the family firm owners go the extra mile for their employees, not only societal reputation will be established but also it will nurture family firms' images to be recognized as the brand name while recruiting high-quality non-family members. Although, there is debate whether to recruit non-family members in family firms (Arijs et al., 2018), however, one thing is obvious which is the intellectual capital needs in family businesses.

Therefore, family firms like any other SMEs are kin to have highly qualified employees to sustain the competitive advantage in the long run. Doing this, the family firms can increase the intellectual capital and generally develop more the human resource pool of highly educated and skillful workers.

According to (Astrachan et al., 2018) the family business brand is constituted of three family brand dimensions, identity view (what values the family managers promote), image view (how the family brand image is represented in the society via intentional or unintentional messages), and reputation view (how these messages are precepted by external stakeholders, which can differentiate them in a positive way from other firms). Each components represents a critical pillar that makes the family firms identified as an important player in the society, which actually builds overall reputation in the industry and the society.

Figure 4.2 shows the TBL's reflection on the family business branding that helps in achieving and sustaining the family firm's competitive advantage. In other words, it suggests that the TBL approach can help the family firms to achieve

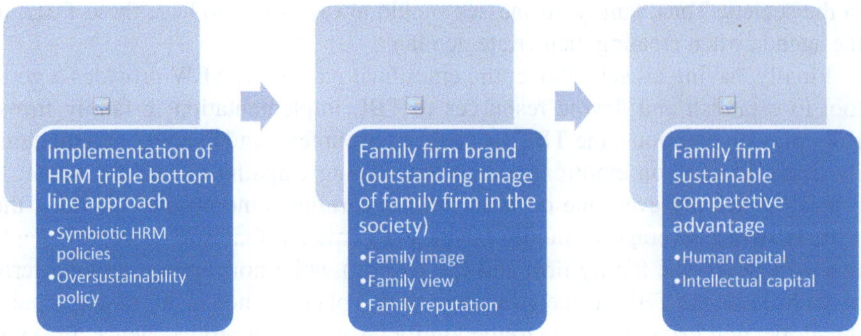

Fig. 4.2 Building competitive advantage based on the TBL approach in HRM. Source: Authors, based on the previous content

long-term sustainability by implementing symbiotic HR policies whereas all human and capital resources will be winners. Once the family firm riches this stage, the family brand is created based on family's image, view, and outstanding reputation (built on the previous two pillars), which further can attract highly qualified employees (family and non-family). Such outstanding human and intellectual capital will ensure the sustainable competitive advantage of the family firm in the market.

4.6 Summary of the Chapter

Traditions are the backbone of family firms and what keeps them going over several generations. The employment relationship model that has been outlined offers a very useful tool for maintaining, advancing, and improving the relationship between a family business and TBL approach without sacrificing the family business's reputation of the business. The model's several levels make it simpler to apply the TBL approach in family businesses from an HRM standpoint.

The background of the model emphasizes the benefit of TBL deployment for HRM. The implications of the model show family businesses how to successfully use the TBL method in human resources while maintaining the family business flexibility and competitiveness in the market.

The final reflection of TBL approach in HRM can be used to develop the family firm's brand. The branding depends on both internal and external factors. The external factors are seen in exceptional products and services (Gallucci et al., 2015) while internal factors could be delineated by cherishing human and physical capitals (Palalić et al., 2021) provided that HRM role will ensure the best practices that will retain and attract new highly qualified employees (intellectual capital). Furthermore, such a branded family firm is ready to compete in the market as well as to retain sustainable competitive advantage.

Case Study[2]: Sustainable HR Initiatives at Suzuki

Suzuki Motor Corporation (hereinafter Suzuki) started as a family business tracing its roots to 1909, when Michio Suzuki founded Suzuki Loom Works in Hamamatsu, Japan. The formal beginning of the company dates to March 1920 when it was officially incorporated as Suzuki Loom Manufacturing Co. Over the years, the company evolved and underwent name changes, becoming Suzuki Motor Co., Ltd. in 1954 and later adopting its current name, Suzuki Motor Corporation, in October 1990. Notably, in 1958, the company adopted the distinctive "S" mark as its corporate emblem. While the company initially focused on loom production, it ceased this manufacturing in 1986.

The Global Expansion

In 1963, U.S. Suzuki Motor Corp., a direct sales subsidiary, was established in Los Angeles. Four years later, in 1967, Thai Suzuki Motor Co., Ltd. was established in Thailand, marking the first motorcycle plant outside Japan. The continuous global expansion never stopped. In 2023, the company's headquarters are still in Japan, but Suzuki operates in over 178 locations globally, producing a diverse range of products, including automobiles, motorcycles, outboard motors, motorized wheelchairs, and electro-senior vehicles. As of March 31, 2023, the company's capital amounted to ¥138,370 million (Japanese Yen) (Fig. 4.3).

Speaking of current Suzuki's organizational structure, it is best presented using its official organizational chart shown in Fig. 4.4.

Suzuki navigating sustainability between SDG and ESG

Suzuki is working hard to achieve excellence in car development, motorcycle development, outboard motors development, and motorized

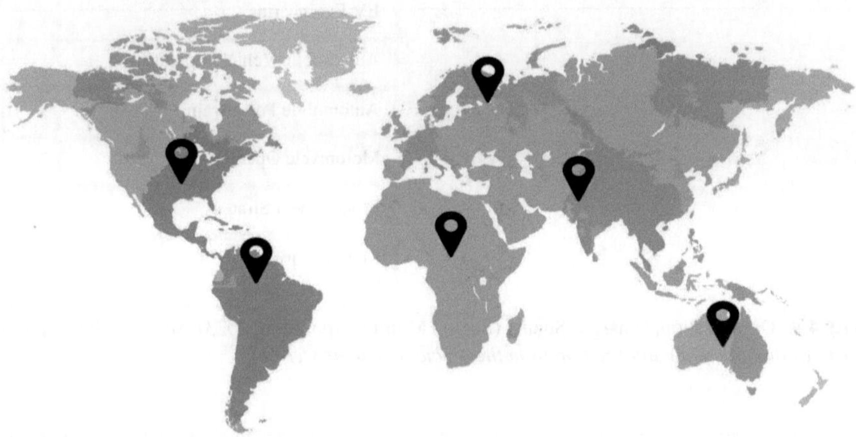

Fig. 4.3 Global operations of Suzuki in 2023 (Suzuki Motor Corporation, 2023). *Source:* Figure developed by authors using information from the official company website

[2] Ensar Mekic (Sultan Qaboos University).

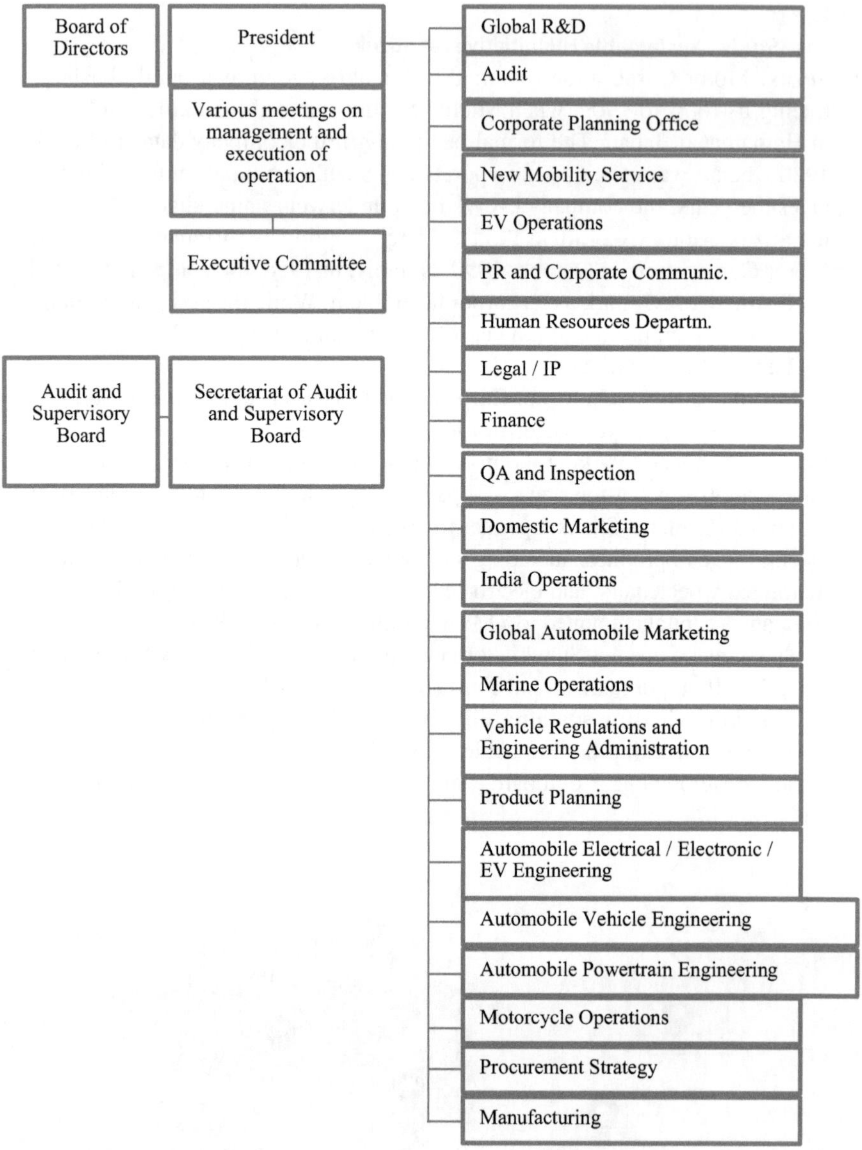

Fig. 4.4 Organizational chart of Suzuki (Suzuki Motor Corporation, 2023). *Source: Figure developed by authors using information from the official company website*

wheelchairs. This however does not prevent the company from being responsible toward people and planet.

Under its mission "*Continue to be an indispensable presence by staying closely attuned to people's lives and providing a means of mobility for local communities*" the Suzuki Motor Corporation never neglected the importance of people, local communities, and their lives within planet. The company's Representative Director and President Toshihiro Suzuki said: *"For FY2030, Suzuki will contribute to the realization of a carbon neutral society and the economic growth of emerging countries such as India, ASEAN, and Africa, with our main business regions, Japan, India, and Europe, as the core."*

Suzuki is committed to sustainability through its policy, aligning with the **United Nations' Sustainable Development Goals (SDGs)** established in 2015. The Suzuki Group actively supports the SDGs, taking responsibility to address relevant issues through its business activities. Suzuki has played a role in developing and popularizing environmentally friendly compact cars, contributing to job creation in emerging countries. The company engages in various community activities, such as promoting traffic safety, providing educational support, participating in disaster relief efforts, and undertaking forest conservation initiatives. Moreover, Suzuki prioritizes human rights, emphasizing key principles like prohibiting discrimination and harassment, preventing forced labor, and opposing child labor. The company values open dialogue and discussion with its employees to uphold these priorities.

When examining Suzuki through the lens of **Environmental, Social, and Governance (ESG)**, particularly focusing on the **environment**, Suzuki has introduced the environmental brand SUZUKI GREEN, which comprises three key categories: SUZUKI GREEN Policy, representing Suzuki's environmental policy; SUZUKI GREEN Technology, showcasing its next-generation eco-friendly technologies; and SUZUKI GREEN Activity, highlighting its environmental initiatives. In addressing climate change, Suzuki has set of ambitious long-term goals, including CO_2 emission reduction from new automobiles, business activities CO2 reduction, air pollution reduction, sustainable water resource use, and waste management.

Suzuki sets ambitious goals not only in environmental matters but also in its **commitment to society**. The company conscientiously fulfills its responsibilities to business partners and employees, prioritizing health management, traffic safety, security, a comfortable working environment, diversity, and fostering healthy employee communication. Additionally, Suzuki demonstrates its commitment to education and culture through the Suzuki Education and Culture Foundation. As of March 31, 2022, the foundation reported the following contributions:

- Scholarships were provided to 524 students, totaling ¥403,380,000.
- Financial assistance to special-needs schools through 75 disbursements, amounting to ¥61,850,000.
- Financial assistance to schools for foreign nationals with 12 disbursements, totaling ¥110,500,000.
- Scholarship aid to Shizuoka University of Art and Culture, supporting 10 scholarships amounting to ¥15,300,000.

In terms of **governance**, Suzuki maintains a clear basic policy. The company aims to earn the trust of shareholders, customers, suppliers, local communities, employees, and other stakeholders through fair and efficient corporate activities. The objective is to contribute further to the international community and ensure sustainable growth and development. To realize this, Suzuki acknowledges the importance of continuous improvement in corporate governance practices. The commitment to enhancing governance reflects the company's dedication to upholding high standards, fostering transparency, and meeting the evolving expectations of its diverse stakeholders.

Sustainability-Oriented R&D at Suzuki

From 2023 to 2030, Suzuki has outlined an ambitious investment plan, earmarking ¥2 trillion (Japanese Yens) for research and development (R&D) expenses. This substantial investment will be allocated to various areas, including carbon neutrality initiatives such as electrification, biogas, and autonomous technologies. Suzuki plans to allocate ¥2.5 trillion (Japanese Yens) for capital expenditures in facilities. This includes the construction of a Battery Electric Vehicle (BEV) battery plant and renewable energy facilities. These capital expenditures signify Suzuki's commitment to infrastructure development that aligns with its goals for carbon neutrality and environmental sustainability. In essence, Suzuki's total investment plan amounts to ¥4.5 trillion (Japanese Yens), reflecting a holistic approach guided by the Triple-Bottom Line (TBL) principles—focusing on people, planet, and profits.

HR Initiatives at Suzuki

At Suzuki, paramount importance is placed on ensuring stable employment for all. The company actively endeavors to enhance working conditions, fostering a healthy and positive work environment. Embodying the team spirit, employees collaborate, support one another, and collectively contribute to society. This collaborative approach extends to both management and staff, unifying to construct a vibrant and innovative company culture. The company is committed to realizing several key objectives:

- Establishment of a safe and healthy workplace.
- The development of a system that fairly evaluates and supports individuals aspiring to achieve higher goals.
- The cultivation of a robust and harmonious employer–employee relationship.

Suzuki's fundamental safety concept comprises key elements, including a robust safety and health control system. This system is manifested through the biannual meetings of the Central Safety and Health Committee, attended by representatives from plants, offices, and labor unions. These sessions play a pivotal role in establishing fundamental policies pertaining to corporate work safety, employee health, and traffic safety. The committee serves as a vital platform for collaborative decision-making to uphold and enhance safety standards across the organization.

Suzuki prioritizes risk assessment activities as proactive measures for risk prevention. The company systematically identifies potential risks in its operations to enhance overall safety. Through these activities, Suzuki aims to improve safety standards by promoting and implementing effective countermeasures to prevent identified risks. This approach underscores the company's commitment to a proactive and preventative stance in safety management.

The company has implemented a Health Declaration, outlining focused initiatives to promote employee well-being:

- Creating a healthy working environment with a strong emphasis on early detection and treatment, fostering a workplace where employees can stay healthy.
- Regular health checks and achieving 100% participation in thorough examinations.
- Preventing lifestyle-related diseases through company's commitment to maintaining and enhancing the working environment by proactively preventing lifestyle-related diseases, particularly among younger employees.
- Improving mental health by regular stress checks to actively address employees' mental health, striving to create a more supportive working environment.
- Promoting physical well-being with efforts directed toward achieving both mental and physical health. The company aims to create opportunities for developing exercise habits, contributing to a healthier workforce.
- Enhancing health literacy by recognizing the importance of daily health maintenance. The company is dedicated to raising health literacy not just for its employees but also for their families.

Suzuki's initiatives for health improvement of their employees as well as indicators to monitor those are presented in Fig. 4.5.

In the realm of career advancement, Suzuki holds the belief that self-improvement is best achieved not merely by adhering to instructions but by voluntarily setting and working toward goals. To facilitate this, the company has instituted a self-assessment system, rotation system, international training program, and foreign language training programs. These initiatives are designed to empower individuals to proactively shape their career paths, fostering a culture of continuous self-improvement within the organization.

In addressing the need to reduce overtime working hours and mitigate the risk of employee burnout, Suzuki has implemented several initiatives. These measures are designed to prevent health issues arising from prolonged working hours. Key initiatives include:

INITIATIVES

- Improve employees' health literacy.
- Increase the rate of receiving thorough examinations after regular health checks.
- Increase the rate of specific health guidance given.
- Address omen's health-related issues
- Reduce smoking rates.
- Health management initiatives for employees assigned overseas.
- Combat COVID-19.

INDICATORS

- Percentage of employees with regular health checks
- Percentage of employees undergoing a thorough examination after regular health checks
- Specific health guidance implementation rate
- Stress check response rate
- Smoking rate

Fig. 4.5 Suzuki's initiatives to improve health of employees

- Strict management of overtime working hours based on total working hours.
- Introduction of a flexible time system to reduce late-night overtime.
- Introduction of interval norms between working hours to ensure rest time within working hours.
- Setting up a working day without overtime, which is a balance between work and life.

In its commitment to supporting work and family balancing, Suzuki has implemented a system aimed at creating a **flexible working environment**. Suzuki is actively fostering awareness of work and family balancing throughout the workplace, striving to cultivate an employee-friendly atmosphere. The company's overarching goal is to create a work environment where individuals can continue to work effectively while maintaining a healthy balance between their professional and personal responsibilities. Suzuki has introduced a **Short Hours System**, specifically helping employees with children in the third grade or younger or those responsible for family members requiring nursing care. The company has implemented a **Leave of Absence System**, acknowledging the diverse needs of its workforce, encompassing both men and women. This system is designed to accommodate employees who, despite possessing the motivation and capability to work, require focused time for childcare or nursing care responsibilities. Employees can take up to 365 days of leave per subject family member. This provision extends beyond standard paid vacations and includes a nursing and medical leave system introduced in April 2015.

In addition to the initiatives mentioned earlier, Suzuki provides the following **welfare and benefits** components such are:

- Dormitory for single employees and housing
- Company-subsidized housing

- Sports facilities
- Employee cafeterias
- Asset-building savings program
- Employee stock ownership program
- Housing loan financing program
- Employee vehicle or family vehicle purchase program

Suzuki is actively committed to promoting **women's participation** in the workforce. The company has prioritized increasing the number of female employees holding job titles from assistant manager and above. In addition to that, Suzuki encourages all employees to utilize paid leave, contributing to the establishment of healthier working environments.

Suzuki is dedicated to establishing a work environment that accommodates **individuals with disabilities**, ensuring they can work comfortably. To facilitate this commitment, the company appoints a specialist in charge of employing people with disabilities, as well as a psychiatric social worker in the Human Resources Department to provide individual consultations periodically.

Discussion questions:

1) What is the role of Suzuki's unique HR initiatives in its sustainability?
2) Considering the global HR challenges, is Suzuki's employee-oriented approach a choice, or a must?
3) How ESG standard helps Suzuki monitor its sustainability?
4) Does Suzuki implement TBL, and if yes, would you correlate its implementation with ESG? Please explain your answer.

Bibliography

Suzuki Motor Corporation. (2023). www.globalsuzuki.com. Retrieved December 15, 2023, from https://www.globalsuzuki.com/ir/library/profile/

Suzuki Motor Corporation. (2023). www.globalsuzuki.com/. Retrieved December 2023, from https://www.globalsuzuki.com/globallinks/

Suzuki Motor Corporation. (2022). Integrated report 2022. Suzuki Motor Corporation.

Suzuki Motor Corporation. (2022). Sustainability report 2022. Suzuki Motor Corporation.

Suzuki Motor Corporation. (2023). Fact Book 2023–2024. Suzuki Motor Corporation.

Suzuki Motor Corporation. (2023). *Environmental initiatives*. Retrieved December 2023, from https://www.globalsuzuki.com/corporate/csr_environment/intro/csr_policy.html

Suzuki Motor Corporation. (2023). *Social*. Retrieved December 2023, from https://www.globalsuzuki.com/corporate/csr_environment/csr/

Suzuki Motor Corporation. (2023). *Governance*. Retrieved December 2023, from https://www.globalsuzuki.com/corporate/csr_environment/governance/
Suzuki Motor Corporation. (2022). *Sustainability policy*. Retrieved December 2023, from https://www.globalsuzuki.com/corporate/csr_environment/intro/csr_policy.html
Suzuki Motor Corporation. (2022). *Activities of Suzuki foundation*. Retrieved December 2023, from https://www.globalsuzuki.com/corporate/csr_environment/csr/foundation.html
Suzuki Motor Corporation. (2023, January 26). *Top Message*. Retrieved December 2023, from https://www.globalsuzuki.com/corporate/message/index.html

Questions for Discussion

- Discuss sustainable human resource management in the family business.
- Analyze and explain the sustainable HRM settings in family business.
- Discuss the importance of sustainable HRM in the family business.
- Evaluate, and discuss the tools and mechanisms of sustainable HRM in family business.

Additional Readings

- Macini, N., Fernandes Rodrigues Alves, M., Oranges Cezarino, L., Bartocci Liboni, L., & Cristina Ferreira Caldana, A. (2020). Beyond money and reputation: Sustainable HRM in Brazilian banks. Employee Relations: *The International Journal*, 44(3), 702–728. https://doi.org/10.1108/er-12-2018-0331
- Mak, A., Sugumar, M., Professor Robin Kramar, D., Cheung, L., Mak, A., & Leung, L. (2014). Confucian thinking and the implications for sustainability in HRM. Asia-Pacific Journal of Business Administration, 6(3), 173–189. https://doi.org/10.1108/apjba-02-2014-0029
- Bush, J. T. (2020). Win-Win-Lose? Sustainable HRM and the promotion of unsustainable employee outcomes. Human Resource Management Review, 30(3). https://doi.org/10.1016/j.hrmr.2018.11.004

Suggested Activities

- Select two firms one family and one non-family business. Make an interview with business owners of each and try to compare the strengths of the TBL approach. In your comparison, which one has the strongest TBL? Identify and discuss reasons that may dictate the intensity of TBL.
- In the above firms, compare and come up with the best HR practices aligned with the TBL approach.
- In groups or individually, find the best practices of sustainable HR. Discuss the differences and shortcomings. Debate the best one.

Keywords and Concepts

- Human resource management
- Employment relationship model
- Quasi-Spot Contract
- Mutual Investment
- Over-investment
- Underinvestment
- Imbalanced 1
- Imbalanced 2
- Balanced 1
- Balanced 2

References

Websites

Arijs, D., Botero, I. C., Michiels, A., & Molly, V. (2018). Family business employer brand: Understanding applicants' perceptions and their job pursuit intentions with samples from the US and Belgium. *Journal of Family Business Strategy, 9*(3), 180–191. https://doi.org/10.1016/j.jfbs.2018.08.005

Astrachan, C. B., Botero, I., Astrachan, J. H., & Prügl, R. (2018). Branding the family firm: A review, integrative framework proposal, and research agenda. *Journal of Family Business Strategy, 9*(1), 3–15. https://doi.org/10.1016/j.jfbs.2018.01.002

Aust, I., Matthews, B., & Muller-Camen, M. (2020). Common good HRM: A paradigm shift in sustainable HRM? *Human Resource Management Review, 30*(3), 100705. https://doi.org/10.1016/j.hrmr.2019.100705

Barringer, B. R., & Ireland, R. D. (2019). *Entrepreneurship: Successfully launching new ventures* (6th ed.). Pearson.

Bush, J. T. (2020). Win-win-lose? Sustainable HRM and the promotion of unsustainable employee outcomes. *Human Resource Management Review, 30*(3), 100676. https://doi.org/10.1016/j.hrmr.2018.11.004

Gallucci, C., Santulli, R., & Calabrò, A. (2015). Does family involvement foster or hinder firm performance? The missing role of family-based branding strategies. *Journal of Family Business Strategy, 6*(3), 155–165. https://doi.org/10.1016/j.jfbs.2015.07.003

Lopez-Cabrales, A., & Valle-Cabrera, R. (2020). Sustainable HRM strategies and employment relationships as drivers of the triple bottom line. *Human Resource Management Review, 30*(3), 100689. https://doi.org/10.1016/j.hrmr.2019.100689

Lumpkin, G. T., & Bacq, S. (2022). Family business, community embeddedness, and civic wealth creation. *Journal of Family Business Strategy, 13*(2), 100469. https://doi.org/10.1016/j.jfbs.2021.100469

Palalić, R., Duraković, B., Ramadani, V., & Ferreira, J. J. M. (2021). Human capital and youth emigration in the "new normal". *Thunderbird International Business Review, 65*(1), 49–63. https://doi.org/10.1002/tie.22250

Palalić, R., & Smajić, H. (2022). Socioemotional wealth (SEW) as the driver of business performance in family businesses in Bosnia and Herzegovina: The mediating role of transformational leadership. *Journal of Family Business Management, 12*(4), 1043–1064. https://doi.org/10.1108/jfbm-07-2021-0067

Podgorodnichenko, N., Edgar, F., & McAndrew, I. (2020). The role of HRM in developing sustainable organizations: Contemporary challenges and contradictions. *Human Resource Management Review, 30*(3), 100685. https://doi.org/10.1016/j.hrmr.2019.04.001

Ramadani, V., & Hoy, F. (2015). Context and uniqueness of family businesses. In L. P. Dana & V. Ramadani (Eds.), *Family businesses in transition economies*. Springer.

Ramadani, V., Memili, E., Chang, E. P., & Palalić, R. (2020). *Entrepreneurial family businesses: Innovation, governance and succession* (1st ed.). Springer.

Rothaermel, F. (2017). *Strategic management* (3rd ed.). Mc Graw Hill Education.

Tsui, A. S., & Wu, J. B. (2005). The new employment relationship versus the mutual investment approach: Implications for human resource management. *Human Resource Management, 44*(2), 115–121. https://doi.org/10.1002/hrm.20052

Zellweger, T. M., Eddleston, K. A., & Kellermanns, F. W. (2010). Exploring the concept of familiness: Introducing family firm identity. *Journal of Family Business Strategy, 1*, 54–63.

Zellweger, T. M., Kellermanns, F. W., Eddleston, K. A., & Memili, E. (2012). Building a family firm image: How family firms capitalize on their family ties. *Journal of Family Business Strategy, 3*, 239–250.

Sustainable Marketing and Public Relations (PR) in Family Businesses

5

Learning Objectives
After reading this chapter, you should be able to:

- Understand marketing and PR activities via the prism of TBL.
- Apply TBL using marketing and PR.
- Address the importance of TBL approach in marketing and PR.
- Understand tools and mechanisms of marketing and PR that will enhance TBL implementation in family businesses.

Profile: Patagonia[1]

Founded by Yvon Chouinard in 1973, Patagonia is an American merchant of outdoor recreation apparel, equipment, and gear such as fleeces, jackets, fast-dry tops and bottoms, and sleeping bags. Over its years of development and growth, Patagonia has expanded its product line and included many other sports, such as skiing, surfing, and mountain biking. As of August 23, 2023, Patagonia has more than 2600 stores in the United States and it operates in 70 countries worldwide. Patagonia's core values and purpose are heavily based on the triple-bottom-line strategy, where they foreground people and the planet along with profit. The organization believes that activism and environmentalism matter as much as promoting antiracism and justice between the employees and customers, creating high-quality products to maintain customers and, in the end, to gain profit.

Digital marketing has proven to have an enormous positive impact on consumer behavior, and companies are encouraged by this fact. By promoting

[1] Authors. Based on bibliography used.

R. Palalić et al., *Sustainable Family Business*, Springer Texts in Business and Economics, https://doi.org/10.1007/978-3-031-57999-8_5

products or services through social media, companies are able to create strong relationships with their customers and build and maintain customer loyalty. Every organization must find a suitable media channel to articulate their attempts to protect the environment and show their engagement in various social causes. Concerning Patagonia, sustainable marketing has been playing a large role in expanding its business, increasing its sales, and creating a good relationship between its business and customers. The various campaigns they have created and promoted through their website and social media have helped them attract loyal customers—those who are willing to afford Patagonia's products in return for a guarantee of taking measures to save the planet.

One of the most influential marketing strategies that Patagonia has implemented since its establishment is their advertisement in the Black Friday edition of *The New York Times* entitled "DON'T BUY THIS JACKET." The advertisement was one full page long and it contained a picture of the jacket and long text with detailed reasoning as to why customers should refrain from purchasing the jacket. Some of the reasons mentioned was excessive number of resources needed for the production of the jacket, specifically "135 liters of water, enough to meet the daily needs of 45 people." Other reasons included were the massive carbon dioxide emission of 20 pounds, and a large amount of post-production waste. Kenji Explains in his article that, despite Patagonia having a significant increase in sales (30% increase) when it was not the goal, it still managed to draw attention to one of the most important and discussed global issues.

Consequently, Patagonia has been producing long-lasting and durable apparel while encouraging its customers not to buy new clothing items unless completely necessary—and if so, customers can bring their jacket to one of the stores and their old clothes can be recycled and used to make new items. Patagonia's strategy of sustainable marketing certainly led to more wokeness within the society, and their campaigns promoting environmentalism helped to raise over $100 million dollars, making them one of the key players in climate philanthropy.

Bibliography

1. https://arc.cct.ie/cgi/viewcontent.cgi?article=1013&context=business, accessed on September 9, 2023.
2. https://sustany.org/local/wp-content/uploads/2020/07/44d158f2-d39d-4577-bb3b-cdbecb756a45.pptx, accessed on September 9, 2023.
3. https://en.wikipedia.org/wiki/Patagonia,_Inc. accessed on September 9, 2023.
4. https://www.patagonia.com/home/, accessed on September 9, 2023.
5. https://www.linkedin.com/pulse/triple-bottom-line-marketing-business-strategy-gretchen-grani/, accessed on September 12, 2023.
6. https://www.patagonia.com/core-values/, accessed on September 13, 2023.

7. https://www.superfolk.com/stories/2022/2/17/yc-interview, accessed on September 13, 2023.
8. https://kentrix.in/how-digital-marketing-affects-consumer-behaviour-exploring-the-impact/#:~:text=Digital%20marketing%20strategies%20such%20as,online%20sales%20and%20customer%20loyalty, accessed on September 15, 2023.
9. https://www.patagonia.com/stories/dont-buy-this-jacket-black-friday-and-the-new-york-times/story-18615.html, accessed on September 15, 2023.
10. https://medium.com/@PabloTurletti/the-triple-bottom-line-a-must-for-sustainable-marketing-af638ce20f02, accessed on September 16, 2023.
11. https://www.researchgate.net/publication/315466760_The_Importance_of_Communication_in_Sustainability_Sustainable_Strategies, accessed on September 16, 2023.
12. https://bettermarketing.pub/dont-buy-this-jacket-patagonia-s-daring-campaign-2b37e145046b, accessed on September 18, 2023.
13. https://hbr.org/2012/11/patagonias-provocative-black-f, accessed on September 18, 2023.
14. https://www.marketingweek.com/case-study-patagonias-dont-buy-this-jacket-campaign/, accessed on September 18, 2023.
15. https://www.nytimes.com/2022/09/14/climate/patagonia-climate-philanthropy-chouinard.html#:~:text=Patagonia%20has%20already%20donated%20%2450,major%20player%20in%20climate%20philanthropy, accessed September 18, 2023.

5.1 Introduction

Marketing plays an important role in firms' long-term sustainability. In marketing books, marketing was regarded as imperative to succeed in the market, because marketing actually informs potential buyers and clients (B2B) on the availability of certain products and services and their qualities, price, and convenient places (4Ps). Philip Kotler in his different book editions has a variety of definitions. Due to changes in the market, there was an evolutionary road regarding marketing definitions by Philip Kotler and his associates. One of the recent definitions is the one that has a very profound meaning that fits all types of firms. They defined marketing as an "art and science that desirably creates a positive long-term profitable relationship with customers, whereas customers play the role that provides the firm's value in return" (Kotler et al., 2020).

Marketing in a family business is also very important. Family businesses regard marketing activities as a very important aspect of their business. They are fully engaged in marketing activities because they know that the "family image" is important (Ramadani et al., 2020). Family business owners are very much convinced that marketing (it used to be old marketing strategies) can establish and can damage the family image in society if marketing campaigns are not appropriately conducted. As they do care about family image in society, they believe if the family name or image is deteriorated then negative consequences will assail on the business. As marketing can have negative consequences for family businesses, it has also its another positive side. This is because marketing can be a very good tool for family business owners to create an affirmative view of the society for the business. Therefore, family firms carefully are alerted and very sensitive to negative distortions in the society about their business name and business per se.

There are times when the image of the family business needs to be protected. One of the roles of Public Relation (PR) office is to make sure that the image of the family name and/or family business is not damaged. Accordingly, PR team is trying to be informed, notified, or alerted on some unusual activities that can negatively affect the family business. In their endeavors, PR offices establish a strategy of safeguarding the family image publicly and responding to clients' or customers' complaints and needs. The PR's role is not only to protect but also to respond customers' needs in a way that can build trust between the family firm and customers. The long-term perspective of this relationship is mandatory and therefore the role of PR is embedded in marketing as its integral part.

As we discussed in the second chapter, sustainable competitive advantage is imperative for family firms' existence, and thus, marketing and PR are orchestrating together to achieve the sustainability of the family firm in regard to the triple-bottom-line approach.

As we introduced in the discussion above, the primary role of marketing and PR in family firms or any other type of firm, is to seek the long-term approach, the triple-bottom-line approach, which will bring benefits in the long run, for the family enterprise, the environment, and people.

The next section in this chapter explores possible ways to implement the triple-bottom-line approach regarding sustainable marketing and PR activities. The aim is to introduce possibilities that can help family enterprises to compete within the new rules of the market game. The market actually imposed other types of stakeholders that must not be neglected, the people and environment.

5.2 Nature of Marketing and PR

Discussing on marketing activities we already tackled a bit of the role of marketing and PR in family firms. The nature of marketing and PR is to act complementarily to help the family business to keep, enhance, improve, and eventually to protect the family business name, image, and reputation in the market. The TBL approach has more positive sides that actually can bring the prosperous growth of family firms. However, it is a challenge at the same time. The first and foremost challenge in the wholistic approach of the TBL model is very sensitive while implementing this approach, because if not appropriately conducted, it may degrade the family business name. The sensitivity is reflected in very tiny and sensitive customers' behaviors, views, attitudes, and reactions toward family firms' actions. Moreover, they may act against the existing family business. Thus, for the "figure father" (Palalić & Smajić, 2022; Ramadani et al., 2020), it is a must to have proper marketing activities that will only enhance the family business' image, not otherwise.

Having in mind the natural posture of marketing and PR, it is unambiguously important to understand the implication as well as application of marketing and PR roles in TBL approach implementation.

Family businesses have advantage over ordinary SMEs in implementing the triple-bottom-line approach because their long-term sustainability is also related to a tradition that is established throughout years of being active in the market and the society. This, the TBL approach, is in fact a booster that can be used to leverage the long-term sustainability of the family firms.

Sustainable marketing is defined as a concept that advocates for the creation of products in a way that meets the demands of all consumers while avoiding any adverse effects on future generations (Ottosson & Parment, 2015). Sinčić Ćorić and her associates (Sinčić Ćorić et al., 2020) aimed to provide a sustainable marketing concept that will boost the performance of startups. They proposed so-called "Sustainable Marketing Orientation" (SMO) composed of "a three-dimensional formative construct comprising strategic integration, societal engagement, and ethical capabilities." Each dimension plays an integral part in the sustainability through marketing activities. This concept is yet to be implemented in startups, SMEs, or family businesses.

Prior to implementation of the TBL, we believe that family firms should establish the following framework (eight steps) that will be useful to accommodate and implement the TBL approach.

These **eight steps** are:

STEP 1: *Design of TBL-sustainable marketing strategy (TBL: full approach).*

STEP 2: *Development of TBL-sustainable products Services (TBL-PROFIT).*

STEP 3: *Design all TBL-sustainable activities toward all stakeholders (Family Firm Members, Employees, Customers, Clients) (TBL-PEOPLE).*

STEP 4: *Development of strong ties with society (TBL: PEOPLE).*

STEP 5: *Development of TBL-sustainable CSR (TBL: PLANET).*

STEP 6: *Development of Green marketing strategy (TBL: PLANET).*

STEP 7: *Development of TBL-sustainable and all-inclusive transparent communication (TBL: PEOPLE) Transparency and reporting are usually related to financial reporting on financial activities in firms (Bhaduri & Ha-Brookshire,* 2011*).*

STEP 8: *Development of TBL-sustainable marketing activities control (TBL: full approach).*

Let us look at each of the steps, its importance for the family businesses, and how family firms can implement them.

5.3 TBL 1: Profit

5.3.1 TBL Strategy Development for Long-Term Sustainability Vision (TBL: Full Approach)

It is very known that most family-owned companies are into long-term strategic planning. It is because of the family business like so-called "longer-term horizon" (Schmieder, 2014) approach. Major reasons behind this approach are that family firms, according to Schmieder, like to be:

- *On safe side in terms of risk.* They do not like to be exposed publicly, like in capital markets because if a firm is listed in the capital market then the borrowings are inevitable, which in the long-term creates a dose of risk which is bigger than the family-owned firm can tolerate.
- *To have time to develop an innovative product/service.* Family business owners like to have plenty of time to plan and develop an innovative product. Family businesses prefer to be on the safe side rather than rush into innovation processes with all the stakeholders they can bring in because the design may have been a mistake. The process itself is not radical but incremental.
- *Lower-profit scale orientation.* This reason is another prevention of not being exposed to big risks, and the family firm owners are not like ordinary SMEs' CEOs preferring quick and big scale revenues in a short time period.
- *Smooth succession between generations.* As the family business is set to be "generational" potential successors of firms need time to learn so that they will be able to take over the family business from his/her "big father" (Ramadani et al., 2020).

The marketing department and PR should work together to develop long-term strategic planning-related 4Ps marketing strategy. The traditional 4Ps marketing should be integrated into the TBL-sustainable marketing in a way that all four parts leverage the idea of keeping an eye on social, environmental, and economic outcomes of new products.

Therefore, the 4Ps should designed as **TBL-4Ps** (Fig. 5.1), like shown below:

1. TBL-1-PRODUCT *(TBL-sustainable products that address positive social needs)*
2. TBL-2-PRICE
3. TBL-3-PLACE
4. TBL-4-PROMOTION

Let us now look at each of these new TBL-sustainable 4Ps and their features that need to be integrated into sustainable (TBL) marketing tools that will enhance the family business sustainability in the long run.

TBL-1- PRODUCT. Traditional marketing tool (product) was mainly aimed at producing quality products or services. As the need of this century changes, this product quality is still imperative, but the quality should address society needs and

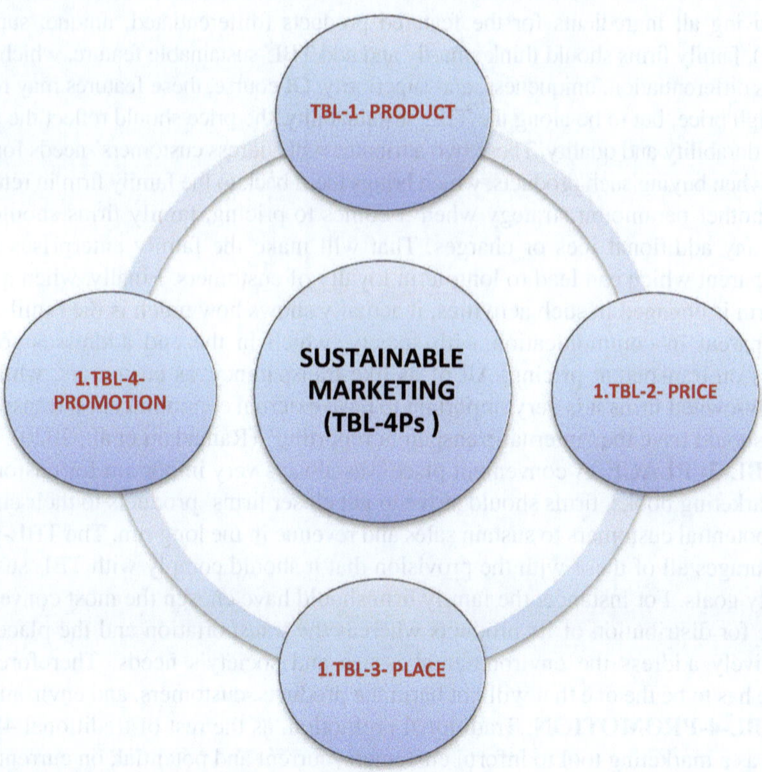

Fig. 5.1 Sustainable marketing: TBL approach

it should be produced or delivered (if service) with environmentally safe features (Dyck & Manchanda, 2021). For instance, family-owned firms are discouraged from producing or providing products and services that are quality but have disastrous features toward environments and health of customers.

Therefore, the *product should address the quality, safety, and reusable feature of products in the future*. Along the same line, these TBL-sustainable products should be used in other social activities, sponsorships, sales shows, etc., that will encourage customers to buy, and at the same time, the family firm shows how much is responsible toward the society, the consumers, and the environment. Finally, the products should not be ended by one of the above activities. Rather there should be an incremental innovation, as discussed earlier in the text. Meaning, every new sales cycle should have a newly designed product with TBL features.

TBL-2- PRICE: Kotler and his associates in their marketing books always stress on the differentiated value that actually brings long-term sustainable revenue. Defining marketing itself as a "long-term profitable relationship" with current customers, Kotler and Keller (2015) actually explained the importance of the proposition value of a product.

Having all ingredients for the featured products (differentiated, unique, superior value), family firms should think smartly and add TBL-sustainable feature, which integrates differentiation, uniqueness, and superiority. Of course, these features may reflect the high price, but to be along the TBL sustainability, the price should reflect the products' durability and quality. These two attributes will address customers' needs for savings when buying such products, which brings them back to the family firm in return.

Another paramount strategy when it comes to pricing, family firms should not hide any additional fees or charges. That will make the family enterprises more transparent which can lead to long-term loyalty of customers. Finally, when a family firm is engaged in such activities, it actually shows how much is the family firm transparent in communication with society, which in the end addresses society issues on transparent pricing. All of us like transparency, as customers, while for family-owned firms it is very important to have external communication transparent as it should have the "internal-transparent reporting" (Ramadani et al., 2020).

TBL-3- PLACE: A convenient place was always very important for customers. In marketing books, firms should strive to get closer firms' products to their current and potential customers to sustain sales and revenue in the long run. The TBL-place encourages all of these with the provision that it should comply with TBL sustainability goals. For instance, the family firm should have chosen the most convenient place for distribution of its products whereas the transportation and the place will positively address the environmental issues and society's needs. Therefore, the place has to be the one that will not harm the products, customers, and environment.

TBL-4-PROMOTION: Traditional promotion, as the rest of traditional 4Ps, is used as a marketing tool to inform customers, current and potential, on current, and new products and services the firm provides. Similarly, the family enterprises will enhance their sustainability via TBL-sustainable promotion. In first place, the family firm should *educate* its customers on TBL's sustainable goals and their effects. With that, the family firm will enhance the sales of TBL-featured products and services because TBL-sustainable products will bring long-sustainability for the society, people and have a positive economic impact on the firm.

Furthermore, the family-owned enterprise should reveal the so-called "*causality*" (Huertas-García et al., 2017; Koschate-Fischer et al., 2012) if customers buy its products. Supporting the firm's TBL-sustainable products, customers will contribute collectively to a sustainable society and environment. At the same time, the family business benefits from all the profits it makes from such sold products. Simply, customers are encouraged and introduced to be a positive cause for the society and the environment. Marketing managers in family enterprises should be aware of the fact that not all customers will respond to their promotional campaigns.

Therefore, the promotion must be *targeted* to work. It means that the target should be the one who supports environmentally or green products that the family firm produces (Reddy et al., 2023). Additionally, to have success in delivering the main message, the message sent should be *appealing* one. Also, the promotional activity should be followed by a very *honest and sincere reporting* on activities to all stakeholders of the family firm, on what the family firm has done in terms of TBL-sustainable products and their positive impact on the planet, people, and profit.

5.4 Economic Perspective of TBL Marketing and PR

Most entrepreneurial enterprises try to achieve sustainable competitive advantage through sustainable strategies that leverage the enterprise's growth and development (Rothaermel, 2017). As discussed in Chap. 2, the competitive advantage is *relative* because it is subject to continuous improvement and advancements of existing and modified products and services. Similarly, the TBL approach to achieve profit in the family firm is based on the design and development of products or services that will address societal and people's needs. The next section explores opportunities that will enhance the family firm's profit through TBL-sustainable products and services.

5.4.1 STEP 2: Development of TBL-Sustainable Products Services (TBL-PROFIT)

The development of TBL-sustainable products is a true challenge. In first place, the family firm must truly understand the TBL approach and adjust the development of new products process according to TBL general policies.

The development process of TBL-sustainable products consists of the following stages (Fig. 5.2) below:

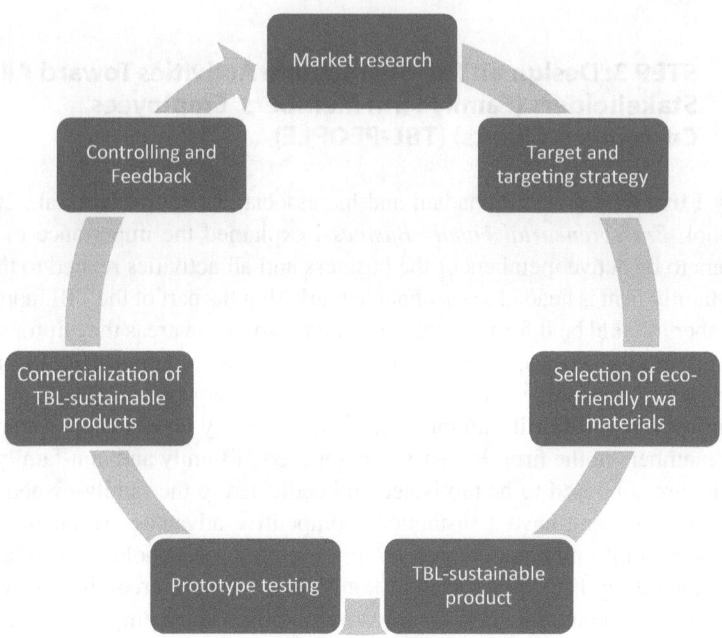

Fig. 5.2 Development process of TBL-sustainable products. *Source:* Authors

These stages are as follows:

- STAGE ONE: *Market research*. It includes gathering all necessary information on market needs that implies the customers' needs who are looking for environmentally safe products.
- STAGE TWO: *Evaluation of identified target group and developing a specific target-driven marketing strategy*. It implies screening of possible target groups that can fit the best the family firm's products.
- STAGE THREE: *Evaluation of alternative eco-friendly materials from which the final products can be made*. Selecting eco-friendly raw materials that will be embedded from the very beginning into TBL-sustainable products. This selection will enhance the durability and quality of the developed products.
- STAGE FOUR: *Developing the concept ofTBL-sustainable prototype*. Production of physical products that are featured with attributes that make green products customers happy and loyal in the long run.
- STAGE FIVE: *Testing the prototype*. This implies testing the demand for the produced TBL-sustainable product.
- STAGE SIX: *Commercialization of TBL-sustainable products*. Real-time sales of the product and real-time feedback on the product life cycle.
- STAGE SEVEN: *Controlling and feedback*. Based on sales and profit of eco-friendly products, the family firm will take corrective actions in order to enhance the existing products and marketing activities to keep the product in the market (target groups).

5.4.2 STEP 3: Design all TBL-Sustainable Activities Toward All Stakeholders (Family Firm Members, Employees, Customers, Clients) (TBL-PEOPLE)

Family Firm Members. Ramadani and his associates (Ramadani et al., 2020) in their book *Entrepreneurial Family Business*, explained the importance of family members to be active members of the business and all activities related to the firm. As the family firm is headed to sustainable marketing (as part of the TBL approach), its members should be informed, educated, and involved in areas they fit for sustainable marketing activities. Each member can play an important role in properly implementing sustainable marketing activities in the long run.

Employees (non-family members). A private entity does not fully employ all family members in the firm. Rather it is composed of family and non-family members who are supposed to be motivated and dedicated to the family-owned firm to prosper it, grow, and have a sustainable competitive advantage (Ramadani et al., 2020). Non-family members represent very important stakeholders who help creation of the family firm values such as family firm's image, credibility as well as a positive societal reputation. In some private enterprises, these employees act as family members because they feel and act as family members. Perhaps, the "big father" (because he created the family firm's values) made them feel this way because he showed respect and appreciation for what they do to the family business. Therefore, in this case the reciprocity really matters. Nonetheless, these employees should be educated and practice the eco-friendly habits and rituals in a way that will contribute to the overall TBL approach in the long term.

Customers and clients. Like employees, customers and clients represent a distinguished stakeholder that should be informed throughout educational practices by the family firm so that they will proudly further promote the TBL agenda for the products the private firm produces. They should see the value in these products, and they will for sure, if the private firm gives some time to impress them by providing all necessary information on eco-friendly goods and services provided. Also, a similar strategy and tactics can be applied toward the family firm's suppliers.

5.5 Societal Impact of Marketing and PR

5.5.1 Development of Strong-Tie with Society: STEP 4 (TBL: PEOPLE)

The legacy of family-owned companies is that they have always had a good relationship with society. The role of the family firm in the society is one of the blocks by which the society is developed, economically and socially. The ancient Greek theory of "Social Contract" (Angus, 2020) could help private enterprises to understand their role in society better. The theory suggests that individuals and groups are fobliged to contribute to socio-economic development by giving up some individual or group freedoms. So do family firms.

The relationship between the family enterprise and society is so natural than any of SMEs and society. It is because for any relationship it takes time, patience, consistency and very honest, immersed, and dedicated effort. As it is a quite-long and consistent effort, establishing a long term and fruitful (strong tie) with the society can be done through the following activities:

- **Open communication with the society** (truly and transparently communicate and show ethical behavior toward the community and society at large).
- **Community engagement** (actively participate in any common-wellbeing activities in the community and society at large).
- **Acting as philantrop in the society** (give up profitable investments to charities like building homes for people in need).
- **Sponsoring for a lifelong education about eco-friendly products and services** (be the first who sponsor the sustainable development of the society by promoting healthy and eco-friendly products and services).
- **Supporting local employment and suppliers** (supporting partners and suppliers is well-known good initiative for the family firm. Additionally, employee veterans or similar categories as a priority, bring back trust to the society which in return gives a lot in return.

The above activities can be used for binding social ties in the community, Ramadani and his associates (Ramadani et al., 2020) explained that these activities, among others, are very important for the family firm to not lose their image in society. Basically, the family firm should be aware of all happenings and developments in the community they operate.

5.6 Environment and Sustainable Marketing and PR

Being socially responsible, family businesses play an important role in socio-economic development of every country. By following the above steps, it is recommended that family firms are acquiring new strategies regarding marketing and PR. Let us now look at the following parts that enable us to understand the overall idea of environment and sustainable marketing and PR.

5.6.1 Development of TBL-Sustainable CSR: STEP-5 (TBL: PLANET)

Building a family image in society is one of the imperatives of the family businesses (Ramadani et al., 2020; Zellweger et al., 2012). This represents a long-term investment, which is durable and constantly creates value for family firms as long as they exist in the market. As we mentioned before, the societal impact of the family firm is a long and constant process that only changes the ways of doing it over time. Corporate Social Responsibility (CSR) in firms generally is discussed by many books and authors emphasizing its importance for their sustainable competitive advantage. Different themes of CSR were recognized and new CSR features were identified that should be emphasized more, like the CSR outcomes in family firms (Stock et al., 2023).

Therefore, it is important that family businesses, as long as they are involved with the society's activities, provide key performance indicators (KPIs) so that they can monitor their activities that positively affect society. Some family enterprises do not like to be exposed so much in the sense that they will reveal what they do for their society, rather they keep a hamble appearance. Often, some family businesses do not communicate their performed societal activities on their websites to their communities. Being discrete and unassertive, family enterprises, however, should track all activities (according to KPIs) they do and proudly, and in a transparent manner share them with their all stakeholders.

5.6.2 Development of Green Marketing Strategy: STEP-6 (TBL: PLANET)

Notion of *green marketing* is accelerated lately, as the importance of triple-bottom-line approach significance enormously increased. **Green marketing** is a new approach to promoting products that are designed and produced in an eco-friendly manner while they have no harm to environment or have a positive impact on our planet.

Because of the awareness of consumers on a critical stage of our planet, eco-friendly products demand increased in recent years (Reddy et al., 2023). Such consumers are also known as LOHAS consumers (Lifestyle of Health and Sustainability). LAHOS consumers (Pícha & Navrátil, 2019) are defined as "consumers are seen as environmentally aware, socially attuned and with a view of the world that takes into account personal, community and planetary outcomes" (p. 233).

Early this year (2023), Starbucks's shareholders agreed on a "planned investment" of over $50 million to "cut its water and waste footprints in half by 2030.". Vice-President of Global Coffee (Social Impact and Sustainability), Michelle Burns said: "At Starbucks, it's our role and responsibility to protect our planet and give back more than we take. These investments will help us make significant strides toward our sustainability commitments while building a business that scales for good and is driven by innovation" (Starbucks, 2023).[2] It is obvious that Starbucks invests in green marketing and eco-friendly packed products. It is has shown publicly that the firm stems to save the environment and inform its customers globally while further educating them on green marketing and eco-friendly products. This news and publicity are well-managed by the Starbucks' Public Relations (PR).

The implication of the role of PR is actually that follows each step the family enterprise takes and reports it, or communicates with community or society at large. If the family-owned business is at the international or global level, then PR communicates with all other branches outside the home country and report those activities to them accordingly. Each of the branches' PR (if exists) reports internally and communicates with the public on such good actions.

5.6.3 Development of TBL-Sustainable and all-Inclusive Transparent Communication: STEP-7 (TBL: PEOPLE)

Transparency and reporting is usually related to financial reporting on financial activities in firms (Bhaduri & Ha-Brookshire, 2011). However, reporting on good actions of the family firm is also regarded as the transparent reporting. In this case, PR has the main role regarding internal and external communication with all stakeholders (shareholders, employees, family members, customers, clients, suppliers, business partners).

PR should have simultaneous communication with each of the stakeholders in a way that provides all necessary information related to TBL marketing strategies

[2] The full story is available on: https://stories.starbucks.com/press/2023/starbucks-commits-over-50-million-to-sustainable-waste-and-water-funds-toward-a-resource-positive-future/.

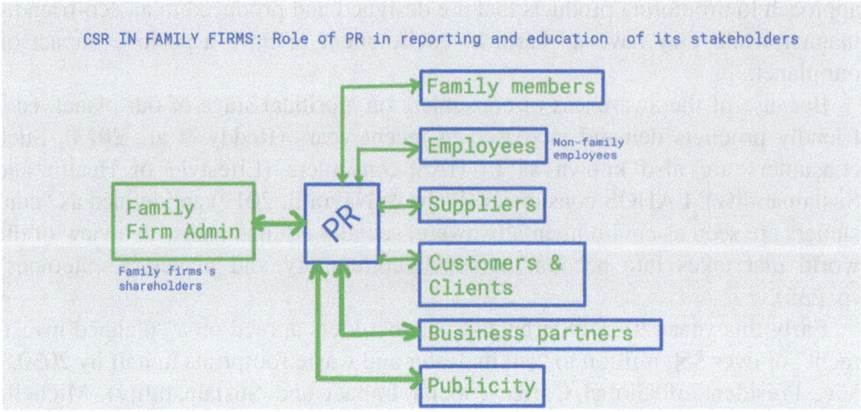

Fig. 5.3 CSR in family firms: Role of PR in reporting and education of its stakeholders. *Source:* Authors

(Fig. 5.3). As the figure shows, PR is the main milestone to communicate, educate, and inform all stakeholders of family enterprise on its TBL marketing initiatives: shareholders, family members, employees, customers & clients, business partners, and publicity (society at large). PR should timely respond to all needs of all stakeholders.

5.6.4 Development of TBL-Sustainable Marketing Activities Control: STEP-8 (TBL: Full Approach)

The last step in the development process of TBL-sustainable products is *controlling*. At this point, family firm collects feedback on the whole process. The feedback provided makes new add-initiatives, corrections of unproved actions, or designing a new strategy that will revise and set up completely new marketing strategies. In other words, the marketing department should collect all key performance indicators at each step and check the overall performance. Then actual performance is compared with planned. All anomalies in actual performance should be revisited and addressed appropriately.

At this stage, each step is compared planned vs. actual. Upon all identified failures, shortcomings and positive sides, a new TBL marketing initiative should be initiated according to controlling reports that address each issue and solution for it (Fig. 5.4)

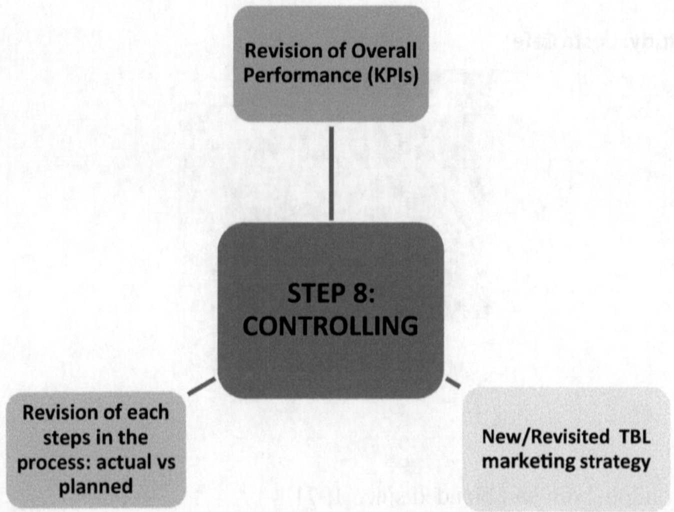

Fig. 5.4 Controling: Step * in the development process of TBL-sustainable products. Source: Authors

5.7 Summary of TBL for Marketing and PR

Triple-bottom-line approach for marketing and PR in family business should be far from "centric" (Dyck et al., 2023) approach. As 4Ps are for decades (Dyck et al., 2019) have been tools for developing smart marketing strategies while PR activities are based on that strategy, however, nowadays for the long-term sustainability it is necessary to engage the community (society and environment) and people (other stakeholders in family firms). Hence, the traditional 4Ps, if the family firms want to be in line with the TBL approach, should be modified as **TBL-4Ps** (Fig. 5.1).

The TBL marketing approach is successful if used in the framework of eight steps (Fig. 5.2). Each step is regarded as a very important pillar in the long run sustainable competitive advantage of family enterprises while using a fully TBL approach.

Case study: Costa Cafe[3]

[4]"London, born and blended since 1971"

"We've 50 years' experience of crafting the finest quality coffee. From revolutionary methods and commitment to quality to unforgettable successes that have made Costa Coffee the Nation's Favourite* coffee shop, our story is as unique as our coffee."

Brief History of Costa

The Costa Coffee story began back in 1971 when Sergio and Bruno arrived in London with a burning desire to make great-tasting coffee a part of everyday life.

As the demand for their coffee increased, transforming a sizable piece of grassland into a brand-new roastery, in 1978, Old Paradise Street sign, transformed a sizable piece of grassland into a brand-new roastery. The Costa Coffee roastery was situated on Lambeth's Old Paradise Street until 2017. This was the location of the new site. From this point on, Sergio and Bruno created and supplied their exclusive blend to some of the city's most prestigious restaurants.

From strength to strength

The first exterior view of the original Costa Coffee location of the Costa Coffee store opened its doors on Vauxhall Bridge Road in 1981. To provide consumers with the finest possible coffee experience, the brothers created this unique establishment where they served espresso and created handmade cappuccinos in porcelain cups, making them the first coffee suppliers in London. Shops soon began to sprout in London's most affluent locales, quickly becoming hotspots for the city's stylish residents.

[3] Authors. Based on bibliography used.

[4] ©Ramo Palalic.

The new home of Costa Coffee

[5]After years of building the brand, they finally opened their doors to a brand-new, cutting-edge roastery in Basildon, Essex, in early 2017. Their roastery, which is among the largest in Europe, enables them to produce 45,000 tonnes of coffee annually, which is a fourfold increase in their capacity to roast coffee compared to our previous location in Lambeth. Not only is their roastery the largest, but it is also among the greenest in the world, and they ethically source all of their raw coffee beans from 100% Rainforest Alliance CertifiedTM farms, which prepares them for delivery to Tilbury Docks, Essex, which is only 14 miles from the Costa Coffee Roaster.

Costa Foundation

The Foundation's mission is "to improve the life chances of boys and girls in coffee growing communities by providing the opportunity of a safe, quality education. Our vision is to eradicate poverty in coffee growing communities." "Our impact so far: Over £20 million spent on more than 100 new schools, over 800 classrooms in 10 countries, giving over 120,000 children access to education in coffee-growing communities across the world."

They claim that they are "different" and give an extra mile being involved in societal activities like the ones below:

- "Building new schools or extending existing ones
- Furnishing the new buildings
- Investing in water supplies
- Providing electrical infrastructure

[5] ©Ramo Palalic.

- School projects that support access to, or enhance, children's education
- Providing computers
- Developing land for families to grow crops
- Building teacher and student accommodation
- Building latrines
- Building science laboratories"

Costa's Sustainability Corner[6]

Sourcing coffee beans sustainably

Costa's fundamental component is coffee, and they think that excellent coffee has the ability to produce positive changes. They are proud to source it sustainably for this reason. They choose their coffee carefully, sourcing 100% of their well-known beans to the high standards set by the Rainforest Alliance. In order to combat the effects of climate change and sustain farmers' livelihoods, they are also looking at new ways to collaborate with them. Their sustainability is based on the three pillars "*now* (100% of our coffee beans in the UK are Rainforest Alliance Certified.), *near* (we'll collaborate with coffee experts to run projects focused on implementing more sustainable coffee farming methods.), and *next* (we'll assess the impact our business has on nature, and set specific targets to reduce it.)."

Costa's future sustainability

[7]The Rainforest Alliance is a global non-profit organization that seeks to normalize good business practices in order to improve the future for both

[6]Costa: Available at https://www.costa.co.uk/sustainability/coffee accessed on 25-12-2023.

[7]©Ramo Palalic.

people and the environment. Its goals are to preserve forests, enhance the standard of living for farmers and farming communities, and assist them in adapting to climate change by collaborating with farmers, businesses, and consumers. Businesses and consumers worldwide recognize the Rainforest Alliance and its small green frog seal as a symbol of social, economic, and environmental sustainability. A wide range of various areas is greatly impacted by Rainforest Alliance certification. These can be regarded as the 3Ps, people, planet, and profit.

Cups and packaging sustainably[8]

[9]They say: "The *packaging* we serve our products in is really important to us, to you, and for the planet. That's why we will continue to prioritise the sustainability of our packaging, including our takeaway cups, lids, cans and pods. Wherever possible we will create circular solutions, considering the entire lifecycle of each piece of packaging – reducing waste, boosting recycling, and using recycled material – moving towards reuse as our ultimate goal. Additionally, they say: "We'll give you a free extra bean if you utilize a reusable cup! This implies that you can receive a free drink on us by simply purchasing five beverages. And the finest part? We don't care which reusable cup you use, so go ahead and use it." This is an incredible way of marketing their sustainability.

They also launched the "*refill campaign*," where they challenged all customers to use the same cup for the new cup of coffee. UK stores are providing free refills to encourage customers to use reusable water bottles rather than single-use plastic ones. They opened 2000 stations for refill across the UK.

Creative cup design, where they designed "iconic takeaway cups." An eco-friendly cup that reduces environmental negative effects. When recycled, the new cups have a 26% reduced carbon footprint than a typical takeout cup. Over 165 million cups were recycled through various places and locations. They are also engaged in *coffee pod recycling*.

---

[8] Costa: Available at https://www.costa.co.uk/sustainability/cups-and-packaging accessed on 25-12-2023.

[9] ©Ramo Palalic.

Costa's Climate Concerns[10]

Costa's *climate* is based on the three pillars "*now* (Our Roastery, opened in 2017, was the first roasting facility to achieve the BREEAM Outstanding sustainability certification)," *near* (We'll halve our emissions per coffee serving by 2030 (from a 2019 baseline))", and *next* (achieve Net Zero emissions, across scopes 1, 2 and 3 by 2040)."

Costa is thrilled to be among the 20 UK businesses that signed the British Retail Consortium's "Climate Roadmap," as a founding member in July 2020. A commitment to combating climate change is part of their strategy, which will help the UK get closer to their goal of achieving Net Zero by the government's 2050 deadline.

Costa introduced its new, challenging carbon reduction goals for its entire worldwide business in March 2022. Costa's first goal is to cut emissions by 50% for each serving of coffee. This is their 2030 aim, which has been accepted by the Science Based Targets Initiative's experts. After this is accomplished, Costa will cut emissions as much as they possibly can before funding the creative initiatives to counteract the few problematic emissions that are still there. This helps them in accomplish their greatest goal, which is to become Net Zero by 2040.

The implementation of this goal will be done through their supply chain, their roastery, stores, and partnerships.

Costa's Community Concerns[11]

Costa is deeply engaged in societal, or *community* activities doing an incredible job for socio-economic development across the globe. Similarly,

[10] Costa: Available at https://www.costa.co.uk/sustainability/climate accessed on 25-12-2023.

[11] Costa: Available at https://www.costa.co.uk/sustainability/communities accessed on 25-12-2023.

they have three pillars: "*now* (in 2022, the Costa Foundation celebrates building 100 schools, changing the life stories of 100,000 children.)," "*near* (we'll aim to raise £3 m for the Costa Foundation each year, across our global markets),", and "*next* (we'll double our impact on society through meaningful engagement with our local communities.)." Moreover, Costa is transparently revealing information on dietary and nutrition of their products they provide worldwide as well as all other r activities they do.[12]

Being present for 53 years now, in 43 countries,[13] it can make a great impact on the people's life, the environment, and get the value in return.

Questions for discussions:

1. Based on the information above, does Costa have a truly sustainable path?
2. The case provided a good base of information. Do you think that the marketing activities could be much better in comparison to what they have done and achieved across the globe?
3. From your perspective, how sustainable marketing should be conducted in Costa, if you are not happy with how it is done till now?
4. What would you advise as a smart strategy that will bring more value to the group, and at the same time that will be sustainable and applicable worldwide? Discuss it with examples.

Bibliography

Costa: Available at https://www.costa.co.uk/sustainability/cups-and-packaging accessed on 25-12-2024

Costa: Available at https://www.costa.co.uk/sustainability/cups-and-packaging accessed on 25-12-2024

Costa: Available at https://www.costa.co.uk/sustainability/climate accessed on 25-12-2024

Costa: Available at https://www.costa.co.uk/sustainability/communities accessed on 25-12-2024

Costa: Available at https://www.costa.co.uk/policies-and-reports accessed on 25-12-2024

Wikipedia: Available at https://en.wikipedia.org/wiki/Costa_Coffee accessed on 25-12-2024

[12] Costa: Available at https://www.costa.co.uk/policies-and-reports accessed on 25-12-2023.

[13] Wikipedia Available at https://en.wikipedia.org/wiki/Costa_Coffee accessed on 25-12-2023.

Questions for Discussion

- Discuss marketing and PR activities via the prism of TBL. Give appropriate examples.
- Explain application of TBL using marketing and PR. Provide an examplary story from the market.
- Discuss the importance of TBL approach in marketing and PR. Provide appropriate examples.
- Discuss tools and mechanisms of marketing and PR that will enhance TBL implementation in family businesses. Provide examples and go beyond the current views provided in the chapter.

Additional Readings

- Bhaduri, G., & Ha-Brookshire, J. E. (2011). Do Transparent Business Practices Pay? Exploration of Transparency and Consumer Purchase Intention. Clothing and Textiles Research Journal, 29(2), 135–149. https://doi.org/10.1177/0887302x11407910
- Dyck, B., & Manchanda, R. V. (2021). Sustainable marketing based on virtue ethics: addressing socio-ecological challenges facing humankind. AMS Review, 11(1–2), 115–132. https://doi.org/10.1007/s13162-020-00184-7
- Gleim, M. R., McCullough, H., Sreen, N., & Pant, L. G. (2023). Is doing right all that matters in sustainability marketing? The role of fit in sustainable marketing strategies. Journal of Retailing and Consumer Services, 70. https://doi.org/10.1016/j.jretconser.2022.103124
- Palalic, R., Ramadani, V., Mariam Gilani, S., Gërguri-Rashiti, S., & Dana, L. P. (2020). Social media and consumer buying behavior decision: what entrepreneurs should know? Management Decision, 59(6), 1249–1270. https://doi.org/10.1108/md-10-2019-1461

Suggested Activities

- Select two coffee shops in your neighborhood and portray their TBL approach. Analyze and give reasons for such an approach.
- Find two family firms in your region and do a little research on sustainable marketing. Identify the pros and cons of the firms.
- Imagine your own company and how it conducts the PR and marketing activities regarding the TBL approach. Then, take a family firm that best practices the TBL approach regarding the PR and compare your imagination and reality. Any comments?
- Identify global players and family firms, who are implementing sustainable marketing. Compare their PR activities. Make judgments on it.

Keywords and Concepts

- Sustainable PR
- Sustainable marketing
- Green marketing
- Community engagement
- Marketing 4Ps
- CSR

References

Angus, P. M. (2020, February 06). What role should a family business play in its community? *Harvard Business Review*, 2020.

Bhaduri, G., & Ha-Brookshire, J. E. (2011). Do transparent business practices pay? Exploration of transparency and consumer purchase intention. *Clothing and Textiles Research Journal, 29*(2), 135–149. https://doi.org/10.1177/0887302x11407910

Dyck, B., & Manchanda, R. V. (2021). Sustainable marketing based on virtue ethics: Aaddressing socio-ecological challenges facing humankind. *AMS Review, 11*(1–2), 115–132. https://doi.org/10.1007/s13162-020-00184-7

Dyck, B., Manchanda, R. V., Vagianos, S., & Bernardin, M. (2023). Sustainable marketing: Aan exploratory study of a sustain-centric, versus profit-centric, approach. *Business and Society Review, 128*(2), 195–216. https://doi.org/10.1111/basr.12314

Dyck, B., Walker, K., & Caza, A. (2019). Antecedents of sustainable organizing: A look at the relationship between organizational culture and the triple bottom line. *Journal of Cleaner Production, 231*, 1235–1247. https://doi.org/10.1016/j.jclepro.2019.05.287

Huertas-García, R., Lengler, J., & Consolación-Segura, C. (2017). Co-branding strategy in cause-related advertising: The fit between brand and cause. *Journal of Product & Brand Management, 26*(2), 135–150. https://doi.org/10.1108/jpbm-07-2015-0939

Koschate-Fischer, N., Stefan, I. V., & Hoyer, W. D. (2012, December). Willingness to pay for cause-related marketing: The impact of donation amount and moderating effects. *Journal of Marketing Research, XLIX*, 910–927.

Kotler, P., Armstrong, G., Harris, L. C., & He, H. (2020). Principles of Marketing: Pearson. For Starbucks, 2023: https://stories.starbucks.com/uploads/2024/02/2023-Starbucks-Global-Impact-Report.pdf (accessed on October, 2023).

Ottosson, M., & Parment, A. (2015). Sustainable marketing: How social, environmental and economic considerations can contribute towards sustainable companies and markets. *Studentlitteratur AB, 1*, 214.

Palalić, R., & Smajić, H. (2022). Socioemotional wealth (SEW) as the driver of business performance in family businesses in Bosnia and Herzegovina: The mediating role of transformational leadership. *Journal of Family Business Management, 12*(4), 1043–1064. https://doi.org/10.1108/jfbm-07-2021-0067

Pícha, K., & Navrátil, J. (2019). The factors of lifestyle of health and sustainability influencing pro-environmental buying behaviour. *Journal of Cleaner Production, 234*, 233–241. https://doi.org/10.1016/j.jclepro.2019.06.072

Ramadani, V., Memili, E., Chang, E. P., & Palalić, R. (2020). *Entrepreneurial family businesses: Innovation, governance and succession* (1st ed.). Springer.

Reddy, K. P., Chandu, V., Srilakshmi, S., Thagaram, E., Sahyaja, C., & Osei, B. (2023). Consumers perception on green marketing towards eco-friendly fast moving consumer goods. *International Journal of Engineering Business Management, 15*, 18479790231170962. https://doi.org/10.1177/18479790231170962

Rothaermel, F. (2017). *Strategic management* (3rd ed.). Mc Graw Hill Education.

Schmieder, J. (2014). *Innovation in the family business succeeding through generations*. Palgrave Macmillan. https://doi.org/10.1057/9781137386243

Sinčić Ćorić, D., Lučić, A., Brečić, R., Šević, A., & Šević, Ž. (2020). An exploration of start-ups' sustainable marketing orientation (SMO). *Industrial Marketing Management, 91*, 176–186. https://doi.org/10.1016/j.indmarman.2020.09.002

Stock, C., Putz, L., Schell, S., & Werner, A. (2023). Corporate social responsibility in family firms: Status and future directions of a research field. *Journal of Business Ethics, 190*(1), 1–61. https://doi.org/10.1007/s10551-023-05382-4

Zellweger, T. M., Kellermanns, F. W., Eddleston, K. A., & Memili, E. (2012). Building a family firm image: How family firms capitalize on their family ties. *Journal of Family Business Strategy, 3*(4), 239–250. https://doi.org/10.1016/j.jfbs.2012.10.001

Sustainable Operations and Supply Chain Management in Family Businesses

6

Learning Objectives
After reading this chapter, you should be able to:

- Define sustainable operation management
- Define sustainable chain management
- Know the three pillars of sustainable supply chain management
- Know the strategies for sustainable operations and supply chain management
- Understand the evolution of sustainable operation management
- Understand the challenges in sustainable operations and supply chain management.

Profile: IKEA
Jesper Brodin, born on November 9, 1968, is a prominent Swedish business executive currently serving as the Chief Executive Officer (CEO) of Ingka Group (IKEA). Throughout his tenure at IKEA, Brodin has fulfilled various positions, including assistant to Ingvar Kamprad, Business Area Manager for Kitchen and Dining, and Managing Director for Range & Supply at Inter IKEA Group.

With Jesper Brodin at the helm of Ingka Group, the entity behind the 80-year-old brand known for its cost-effective and visually appealing furniture (though not explicitly intended as family heirlooms), sustainability has taken precedence. This strategic shift has coincided with the company achieving record revenues.

IKEA holds a firm belief that every action we take carries the responsibility to generate a positive impact on people, society, and the environment.

R. Palalić et al., *Sustainable Family Business*, Springer Texts in Business and Economics, https://doi.org/10.1007/978-3-031-57999-8_6

However, achieving this objective hinges upon their business partners sharing their vision and dedication to sustainability. This is where IWAY, the IKEA code of conduct for suppliers, plays a pivotal role. IWAY embodies the IKEA approach to responsibly sourcing products, services, materials, and components. It lays out explicit expectations and operational guidelines concerning environmental, social, and labor conditions, as well as animal welfare. All suppliers and service providers must collaborate with IKEA.

Having spent over two decades working in accordance with IWAY, IKEA has developed a credible and robust system. This system serves as their cornerstone, allowing them to support, challenge, and collaborate actively with our suppliers.

Through IWAY, IKEA makes an impact in four primary areas: Fostering positive environmental effects; Ensuring dignified and meaningful employment; Upholding children's rights; and Enhancing animal welfare throughout the IKEA value chain.

IWAY is rooted in internationally recognized standards and principles governing human rights, environmental preservation, and worker safety and health. It aligns with our own IKEA values and ensures compliance with legal requirements. It stands as the bedrock of IKEA's broader sustainability efforts, involving both direct suppliers and subcontractors throughout the entire value chain.

IWAY is structured around ten principles that encapsulate IKEA's perspectives on various environmental and social issues: Supported by effective routines and open dialogue; Conducting business lawfully and with integrity; Protecting children and promoting opportunities for work, education, and family life; Respecting fundamental labor rights; Ensuring responsible payment, time off work, and opportunities for skill development for workers; Safeguarding workers' health and safety; Ensuring suitable working and living conditions; Protecting the planet; Managing resources sustainably and in a circular manner, including water and waste; and Ensuring decent lives for animals.

The IWAY principles draw from globally recognized standards and principles, such as the UN Guiding Principles on Business and Human Rights, the UN Global Compact's ten Principles, the UN Sustainable Development Goals, and the ILO Centenary Declaration for the Future of Work, among others.

At the core of IWAY is a collaborative approach to continually evolving the framework in conjunction with our suppliers. IKEA aims to collectively achieve responsible sourcing and cultivate a more sustainable value chain. While IKEA defines minimum mandatory requirements, they actively encourage their suppliers to surpass these standards. To facilitate this, IKEA introduced a staircase model for IWAY requirements, encouraging suppliers to focus on continuous improvement and development in IWAY-related aspects.

Source: Authors. Based on https://www.ikea.com/global/en/our-business/people-planet/building-a-better-business-with-iway/; https://time.com/6284013/ikea-ceo-jesper-brodin-interview-leadership/

6.1 Introduction

This chapter deals with sustainable operation management and sustainable chain management by bringing more insight into these topics and the relationship of sustainability with operation management and supply chain management. The chapter is related to three foundational pillars of sustainable supply chain management (SSCM): environmental sustainability, social responsibility, and economic efficiency which includes dimensions such as resource efficiency, emission reduction, ethical labor practices, and economic considerations, which offers a comprehensive analysis of the integral components collectively defining a sustainable supply chain. Advancing in its exploration, the discourse navigates through key strategies inherent in sustainable operations and supply chain management (SOSCM). It also discusses green procurement, lean and efficient processes, energy efficiency, and waste reduction explaining the benefits that family businesses could have by incorporating the new approaches. It addresses the absence of standardized metrics to realize tangible advantages such as cost savings and risk mitigation, this discussion encapsulates the multifaceted aspects of sustainable practices within the operational and supply chain domains.

6.2 Sustainable Operational Management and Supply Chain Dynamics: Foundational Perspectives

Scientists and researchers after careful collection of data all around the world have proved that the world faces severe changes with an expected 2 to 4-scale rise in global average temperature by the year 2100 and that 30–40% of the species could be extinct (Gupta & Palsule-Desai, 2011). Environmental concerns have been a longstanding preoccupation for humanity, garnering increased attention due to the adverse impacts resulting from organizational activities aimed at enhancing profitability (Suki et al., 2023). This heightened awareness has prompted extensive research in the realm of sustainability, exploring various dimensions under the overarching umbrella of "green" (Chen et al., 2022). Significantly influencing this trajectory are key initiatives such as the Sustainable Development Goals Agenda 2030, the Green Deal, and the Paris Agreement, which have spurred researchers to delve into and contribute to the field of sustainable development (Sepasi et al., 2021). The primary objective of sustainability is to discover innovative approaches to meet present needs without compromising the ability of future generations to fulfill their own requirements (Rexhepi et al., 2023).

Embracing sustainability by incorporating sustainable operation management and supply chain management in family businesses not only aligns with the broader societal goal of preserving the environment but also becomes a strategic choice for ensuring the longevity and competitiveness of the family business within the market. The incorporation of sustainable supply chain management and sustainable operation management in family business represents a conscious effort to intertwine financial decisions with ecological responsibility, thus contributing to the larger framework of sustainable development (Kajtazi et al., 2023). Sustainable operations management (SOM) potentially plays a vital role in contributing to solutions for the complex sustainability challenges faced by many organizations, for example, reducing food waste (Dora et al., 2020). Due to limited natural resources, climate change, and growing social inequality, there is a significant focus on the concept of sustainable operations management. Organizations are reassessing their operational practices to address the three dimensions of sustainability: planet, people, and profit. Despite ongoing efforts for over three decades, the field of sustainable operations management is still in its early stages. Sustainability has been of interest to researchers for many years, starting from Gupta which have discussed the need for aligning environmental strategy with operations strategy and the proposed term "environmental operations management" (EOM), and Sarkis (2016) who focused on tools such as: design for environment (DOE), green supply chains, total quality environmental management (TQEM), and reverse logistics (Tayal et al., 2017). Yet, the most significant contribution to the nascent field of sustainable operations management (SOM) came from Kleindorfer et al., who delineated the scope of operations management revolving around the three Ps (planet, people, and profit) across three domains: (1) Green product and process development, (2) Lean and green operations management, and (3) Remanufacturing and closed-loop supply chains. (Tayal et al., 2017). However, many researchers have replaced the term SOM with a more comprehensive term "sustainable supply chain management (SSCM)" (Tayal et al., 2017). SOM also encompasses topics such as sustainable and green supply chains, green procurement, and reverse logistics (RL) (Gunasekaran & Irani, 2014).

Supply chain management (SCM) is defined as the "integration of business processes from end users through original suppliers that provides products, services, and information that add value for customers" (Cooper et al., 1997, p. 2). most sustainable supply chain management (SSCM) definitions incorporate a set of managerial practices that include all of the following (Gupta & Palsule-Desai, 2011):

- Environmental impact as an imperative.
- Consideration of all stages across the entire value chain for each product.
- A multi-disciplinary perspective, encompassing the entire product life-cycle.

Analyzing both definitions reveals common overarching themes related to environmental sustainability (Gupta & Palsule-Desai, 2011):

- *Firstly,* companies should consider the environmental impact of their activities as an intrinsic part of decision-making, rather than merely as a constraint imposed by government regulations, social pressure, or a trend to exploit for a "green" image.
- *Secondly,* companies need to focus on the environmental impact throughout the entire value chain, extending to suppliers, distributors, partners, and customers.

- *Thirdly,* the perspective on sustainability within companies should go beyond a narrow functional view and embrace a broader perspective that integrates issues, problems, and solutions across functional boundaries.

Organizations such as Ashoka advocate for the concept of "Full Economic Citizenship," fostering collaborations among civil society organizations, financial institutions, and policymakers to scale up initiatives addressing the challenges of the century. The shift from extractive economic designs to generative designs necessitates both private innovation and government guidance. Sustainable operations, at its core, form the basis of a generative economy, where economic activities align with their original purpose of meeting human needs. Social finance approaches, by creating incentives for alignment, facilitate a paradigm shift that represents a profound yet often overlooked transformation in global ownership dynamics (Myers and Conte, 2013, p. 6).

The proliferation of these models collectively lays the foundation for a novel economy, bridging the gap between profit making and contributing to the common good. This holistic approach has the potential to impact the middle class by offering affordable products and contributing to the collective well-being of a larger segment of the population (Rexhepi et al., 2023). Implementation of Sustainable Operation Management efficiency in Family Businesses is related to the sustainable business model that they use. The sustainable business model is defined as the alignment of a business with its environment and response to environmental changes and sustainable development (Kajtazi et al., 2023).

6.3 The Imperative of Sustainable Operations and Supply Chain Management

The imperative of sustainability has evolved beyond a mere facet of corporate responsibility to attain the status of a strategic necessity in the realms of business and industry. It appears to be a cornerstone of increasing the responsibility of business organizations. The objective of sustainability is to suggest new ways how to reduce the negative environmental impacts, help conserve resources as well as ensure that business processes are behaving ethically and that they are doing their best to ass less a possible negative impact on the environment. Reducing greenhouse gas emissions, adopting renewable energy sources, and implementing energy-efficient practices appear to be essential components of sustainable operation and supply chain management (Seuring & Müller, 2008). On the other side, the main aim of supply chain management is a pivotal aspect of business operations, which manages the flow of goods, services, and information starting from suppliers to end consumers. On the other side, the main goal of sustainable supply chain management is to extend these practices and to ensure that environmental, social, and ethical considerations are met while this process happens.

The nascent body of literature on sustainable operations and supply chains reveals that researchers have adopted various focal points and methodological approaches in their investigations. A notable area of inquiry involves exploring how managers endeavor to incorporate sustainability principles into their business models, as evidenced by the work of Bocken et al. (2014). Additionally, scholars have undertaken initiatives to develop sustainable logistic management practices (Epstein, 2008) and have documented the intricate nature of sustainable supply chain practices. Taticchi et al. (2013) conducted seminal work, providing an overview of the state of the art in the field and outlining a research agenda, with a particular emphasis on the role of governance mechanisms in promoting sustainable supply chain management.

Contributing to the conceptual foundation of sustainable supply chain management analysis, Seuring and Muller (2008) presented a framework, while Walker and Jones (2012) directed their research toward comprehending internal and external barriers, as well as enablers, of sustainable supply chain management within the retail industry. Another strand of research delves into the nexus between sustainable supply chain management practices and organizational performance (Seuring and Gold, 2013). The literature, however, presents inconclusive findings in this regard. While some studies affirm a direct and positive correlation between sustainable supply chain management practices and organizational performance (Lai and Wong, 2012; Zhu et al., 2007), others report a negative relationship. Yet, studies are highlighting the importance of considering moderating and mediating variables for a nuanced understanding of these relationships.

A notable limitation in many of these studies is their reliance on surveys to examine managerial perceptions of organizational performance, instead of utilizing more concrete and objective organizational performance data. Despite the availability of such data from public sources like firms' sustainability reports, CSR reports, and general annual reports (e.g., Green et al., 2012; Rahdari et al., 2020), prior research appears to underutilize these sources. Addressing this gap in the literature by incorporating more robust and objective performance data could enhance the rigor and validity of future research in the field.

According to Sarkis (2016), sustainable operations and supply chains have a major impact in finding ways to reduce resource consumption and by that minimize the environmental impact. They also include ethical labor practices like having fair wages, working conditions, and respecting business human rights (Carter & Easton, 2011). Most of them are given as guidelines known as UN Guiding Principles. Having these guidelines who are gaining more attention among employees, especially in the EU where in most countries these requirements are becoming imperative. Sustainable operations as sustainability relate also to projects in local communities that have a positive impact on society such as philanthropy, cooperation in social events, and giving to society (Seuring & Müller, 2008). Failure to meet this imperative can lead to legal repercussions and reputation damage to organizations. But not just that more research is pointing to the fact that consumers are paying more attention to the social behavior of enterprises and organizations that fail to achieve this might see a loss in consumers and profitability (Graafland et al., 2014). However, sustainable operations and supply chain management face many difficulties in implementing their principles, as it requires many changes in their processes, organizational culture, and overall infrastructure. This sometimes could be a very expensive process and requires time, also managers might face high resistance to change as well as increased costs of production or distribution (Carter & Easton, 2011).

But most of the obstacles can be overcome if organizations prepare themselves and develop tailored sustainable strategies. Sustainable operations cannot be overstated, it is driven by environmental, social, and economic factors that demand responsible and ethical business practices. Sustainable supply chain management and operation management recently have emerged mainly as a need of time in response to growing environmental and social concerns. The integration of sustainability principles into supply chain operations has transitioned from a trend to a necessity, shaping the future of business practices. These criteria are no longer an option, it is a necessity for organizations in order to address global challenges and ensure the long-term success of the organizations. Embracing sustainable operations could be a strategic move also, it can secure the organization a more prosperous, and environmentally sustainable future.

6.4 The Three Pillars of Sustainable Supply Chain Management

6.4.1 Environmental Sustainability

In recent times, the necessity for businesses to actively engage with environmental sustainability issues in their supply chains has grown increasingly conspicuous. The pursuit of environmental sustainability in supply chain management entails the amalgamation of eco-friendly practices and the mitigation of detrimental environmental effects throughout the lifecycle of the supply chain. This encompasses diverse elements, including conscientious sourcing, energy-efficient transportation, waste reduction, and the embrace of circular economy principles (Carter & Rogers, 2008; Srivastava, 2007). Environmental sustainability places emphasis on two primary dimensions:

- ***Resource efficiency:*** This aspect involves the optimization of processes, waste minimization, and the reduction of the environmental footprint (Sarkis, 2016). The objective is to conserve and prudently utilize natural resources, ensuring their sustained availability for future generations.
- ***Emission reduction:*** In addressing the imperative of combating climate change, Sustainable supply chain management (SSCM) directs its efforts toward diminishing greenhouse gas emissions. This is achieved through the implementation of energy-efficient transportation systems, the adoption of sustainable packaging

practices, and the cultivation of responsible sourcing strategies (Seuring & Müller, 2008). By focusing on emission reduction, SSCM contributes to mitigating the adverse impacts of climate change and aligning supply chain practices with broader environmental objectives.

While the recognition of the significance of environmental sustainability is widespread, the effective implementation of sustainable practices within supply chains is confronted with a myriad of challenges. Issues such as the absence of standardized metrics, the intricate nature of global supply networks, and the resistance to change within organizational structures pose formidable obstacles that businesses must navigate to achieve substantive progress (Seuring & Müller, 2008; Zhu et al., 2008).

Despite these challenges, a multitude of best practices and innovative strategies have surfaced to promote environmental sustainability in supply chains. These encompass the integration of green technologies, collaborative efforts with suppliers to establish sustainability standards, and the integration of life cycle assessments into product design and sourcing decisions (Sarkis, 2012; Schultmann & Sun, 2019).

As the global emphasis on environmental sustainability intensifies, the trajectory of supply chain management is positioned for further evolution. Future avenues of research encompass the exploration of advanced technologies such as blockchain for enhanced supply chain transparency, the examination of the role of regulatory frameworks in propelling sustainability initiatives, and the ongoing development of standardized metrics for the measurement of environmental performance (Mollenkopf et al., 2010; Govindan, 2018).

6.4.2 Social Responsibility

In the contemporary global business landscape, marked by increasing interconnectivity and heightened social consciousness, the incorporation of social responsibility into supply chain and operations management has attained paramount importance. The theoretical underpinnings of social responsibility in this context draw from foundational frameworks such as the Triple-Bottom Line (Elkington, 1998) and Stakeholder Theory (Freeman, 1984). These frameworks emphasize the intricate interplay between economic, social, and environmental factors in decision-making processes, providing organizations with a structured approach to assess their societal and stakeholder impacts. Beyond mere philanthropy, social responsibility extends into critical domains:

- ***Ethical labor practices:*** Within the realm of Sustainable Supply Chain Management (SSCM), ethical labor practices are deemed non-negotiable. Ensuring fair wages, safe working conditions, and the protection of human rights is imperative (Carter & Easton, 2011).
- ***United Nations Guiding Principles on Business and Human Rights (UNGPs):*** Enacted by the UN Human Rights Council in 2011, the UNGPs articulate three

pillars, encompassing the state's duty to protect human rights, corporate responsibility to respect human rights, and the necessity for effective remedies for human rights abuse victims. These principles underscore businesses' critical role in upholding human rights across operations, supply chains, and business relationships, setting a global standard for integrating human rights considerations into policies and practices.

- ***Community engagement:*** SSCM extends to community engagement through corporate social responsibility initiatives, wherein businesses participate in philanthropy, support local communities, and collaborate with stakeholders (Seuring & Müller, 2008).

The integration of social responsibility faces notable challenges, including supply chain transparency, ethical sourcing, and the enforcement of fair labor practices, demanding intricate navigation by organizations committed to upholding their social responsibility mandates (Carter & Rogers, 2008; Seuring & Müller, 2008). Social responsibility, therefore, stands as an integral facet of contemporary supply chain and operations management, guiding businesses in navigating the complexities of a globalized world and fostering ethical and sustainable practices. However, the literature review indicates that Sustainable Supply Chain Management (SSCM) models predominantly focus on deterministic approaches and the integration of environmental aspects of sustainability while neglecting stochastic modeling techniques and the consideration of social factors (Brandenburg & Rebs, 2015).

6.4.3 Economic Efficiency

The central role of economic considerations in the sustainability paradigm aims to provide a nuanced comprehension of the dynamic interaction between economic efficiency and sustainability within contemporary supply chain frameworks. The imperative for businesses to pursue economic efficiency in sustainable supply chain management (SSCM) becomes evident in an era marked by heightened environmental awareness and social responsibility. Achieving economic efficiency in SSCM necessitates the optimization of processes, waste minimization, and enhanced resource utilization. This encompasses the adoption of lean manufacturing practices, efficient logistics and transportation strategies, and the integration of advanced technologies such as blockchain to augment transparency and efficiency.

The theoretical foundations of economic efficiency in SSCM draw from established frameworks, notably:

- ***Transaction Cost Economics (TCE) and the Resource-Based View (RBV).*** TCE provides insights into minimizing transaction costs, establishing the economic imperative of SSCM for cost reduction and heightened efficiency. Sustainability practices within SSCM lead to waste reduction, resource conservation, and ultimately, financial benefits (Sarkis, 2016).

– ***The RBV underscores the strategic deployment of resources for sustained competitive advantage.*** Market demand for sustainability-driven products and services compels businesses to adopt SSCM, critical for maintaining competitiveness and market share (Graafland et al., 2014).

These frameworks offer a theoretical lens through which organizations can analyze and optimize their economic performance while concurrently addressing environmental and social considerations.

However, the pursuit of economic efficiency in alignment with sustainable goals is not without challenges. The intricate task of balancing economic considerations with environmental and social aspects demands strategic decision-making. Implementing SSCM poses formidable challenges, requiring comprehensive changes in processes, supplier relationships, and organizational culture. Key challenges encompass resistance to change, initial investment costs, and the need for standardized sustainability metrics (Carter & Easton, 2011). Moreover, addressing challenges involves investing in sustainable technologies, adapting to changing consumer preferences, and navigating potential trade-offs between economic efficiency and other dimensions of sustainability. As businesses increasingly recognize the interconnectedness of economic efficiency and sustainability, ongoing research and industry advancements will shape the future trajectory of economic considerations within the broader context of sustainable supply chain management.

6.5 Strategies for Sustainable Operations and Supply Chain Management

6.5.1 Green Procurement

In most cases, sustainability in SCM is mainly triggered externally by regulatory pressures and incentives from governments, customer markets, and other stakeholders but much scientific research has proven improved performance of SSCM practices, or some positively affect only the environmental performance of a firm but not its economic or operational performance (Brandenburg & Rebs, 2015). Adopting environmentally sustainable suppliers and products represents a foundational step toward the overarching goal of sustainability. Businesses can establish criteria for suppliers that prioritize eco-friendly production, ethical labor practices, and minimized carbon footprints. This concept, often termed Green Procurement, constitutes a critical element within the framework of sustainable supply chain management (SSCM), playing a pivotal role in mitigating the environmental impact of supply chain operations.

Green procurement, synonymous with environmentally responsible procurement, involves acquiring goods and services with a reduced environmental impact compared to traditional alternatives (Murphy & Poist, 2012). This encompasses sourcing products and services that are eco-friendly, energy-efficient, and ethically and sustainably produced, significantly diminishing the environmental footprint of supply chain activities. By opting for eco-friendly alternatives, organizations contribute to resource conservation and a decrease in greenhouse gas emissions (Sarkis, 2016).

Several principles underpin Green Procurement:

- ***Life-cycle assessment:*** Conducting life-cycle assessments is integral to evaluating the environmental impacts of products and services throughout their entire life cycle (Murphy & Poist, 2012). This comprehensive approach ensures that procurement decisions are based on full environmental transparency.
- ***Ethical sourcing:*** incorporates assessing the social and environmental responsibility of suppliers, considering factors such as labor practices and environmental standards if they have been established in their organizations.
- ***Energy efficiency:*** Green procurement emphasizes the selection of energy-efficient products and services with minimal carbon footprints, thereby reducing energy consumption and greenhouse gas emissions (Sarkis, 2016).

Green procurement encounters challenges like increased costs, internal resistance to change, and supplier collaboration. Overcoming these challenges demands a strategic integration of SSCM fostering supplier collaboration and educating stakeholders about the enduring benefits of environmentally responsible procurement.

6.5.2 Lean and Efficient Processes

Sustainability often correlates closely with efficiency, a connection not always immediately evident to businesses. The optimization of processes to reduce waste and unnecessary resource consumption not only results in cost reduction but also contributes to a diminished environmental impact. Organizations must underscore the significance of attaining operational efficiency while concurrently diminishing environmental repercussions and upholding social responsibility within the supply chain. The amalgamation of lean practices with sustainability principles emerges as pivotal for businesses endeavoring to excel in the contemporary global market.

While lean and efficient processes in sustainable supply chain management present a potent solution for businesses seeking operational excellence while mitigating their environmental and social impacts, the integration of these principles into supply chain management poses challenges. These challenges include the necessity for cultural transformation, resistance to change, and the complexity of managing multiple objectives (Sarkis & Dhavale, 2015). A successful implementation strategy entails aligning lean and sustainability goals, fostering a culture of continuous improvement, and educating employees and stakeholders on the benefits of this integrated approach. The integration of lean and sustainability principles aligns with the growing trend of eco-conscious consumers and the imperative for businesses to meet global sustainability goals. In this context, lean practices emerge as a crucial tool for achieving sustainable supply chain efficiency.

6.5.3 Energy Efficiency and Renewable Energy

The integration of energy efficiency and renewable energy sources into supply chain operations is imperative for achieving sustainability goals within the framework of sustainable supply chain management (SSCM). Companies are increasingly recognizing the need to invest in energy-efficient technologies and renewable energy sources, driven not only by the imperative to reduce their environmental footprint but also in response to recent energy crises, notably those stemming from the COVID-19 pandemic and the conflict in Ukraine, which have led to substantial increases in energy prices. Organizations are acknowledging the pivotal role of energy management in SSCM, highlighting the importance of reducing carbon emissions and minimizing environmental impact. Energy efficiency and the adoption of renewable energy sources represent primary objectives for businesses aiming to align with sustainability goals while optimizing supply chain processes. The integration of Energy Efficiency and Renewable Energy encompasses several key dimensions:

- ***Optimized transportation:*** Energy-efficient transportation modes, route optimization, and load consolidation stand as essential components of energy-efficient supply chain operations. These practices contribute to the reduction of fuel consumption and carbon emissions (Sarkis, 2016).
- ***Renewable energy adoption:*** The integration of solar and wind energy sources into supply chain facilities provides a sustainable and cost-effective means of energy generation (Bazilian et al., 2011). Surplus energy can be returned to the grid, further reducing operational costs.
- ***Carbon emission reduction:*** A central impetus for integrating energy efficiency and renewable energy lies in the reduction of carbon emissions and better control (Goli & Mohammadi, 2022). The adoption of renewable energy sources facilitates the minimization of fossil fuel consumption, a primary contributor to emissions, aligning with global sustainability objectives and climate change mitigation efforts (Bazilian et al., 2011).
- ***Sustainable procurement:*** Procuring energy from renewable sources, such as through green power purchase agreements, contributes to diminishing the carbon footprint of supply chain operations (Sarkis & Dhavale, 2015).

Implementing energy efficiency and renewable energy practices in SSCM poses challenges, including initial investment costs, regulatory barriers, and cultural resistance, which vary across countries. A successful implementation strategy involves aligning energy management goals with sustainability objectives, fostering a culture of energy consciousness, and investing in energy-efficient technologies and infrastructure. Businesses that embrace these principles are better positioned to achieve their sustainability goals, minimize their carbon footprint, and realize cost savings. Energy management emerges not only as an environmentally responsible initiative but also as an economically beneficial and strategic component of supply chain sustainability in the modern era.

6.5.4 Waste Reduction and Recycling

The concerted effort to minimize waste and advocate for recycling programs represents a substantial contribution to sustainability initiatives. Strategies such as reducing the use of disposable materials, initiating recycling programs, and embracing circular economy principles play crucial roles in this endeavor. Within the context of sustainable supply chain management (SSCM), waste reduction and recycling emerge as integral components with pivotal significance in attaining sustainability objectives throughout the supply chain. Minimizing waste is instrumental in conserving resources, and curtailing the unnecessary consumption of materials and energy (Sarkis, 2016). This approach aligns with global sustainability goals, ensuring efficient and judicious resource utilization. Emphasizing the importance of waste minimization and recycling underscores the essential role of environmentally responsible practices in contemporary supply chain operations. Improving the efficiency and sustainability of production processes enables the shift toward sustainable production and consumption resulting in reduced dependency on raw materials, and optimal resource use and recycling (Yazan & Fraccascia, 2020).

Waste management practices, encompassing recycling and waste-to-energy initiatives, effectively diminish the environmental impact linked to landfills and incineration (Braungart & McDonough, 2002). This reduction in carbon emissions contributes to climate change mitigation. Various waste reduction practices within Sustainable Supply Chain Management (SSCM) can be implemented:

- ***Optimized packaging:*** Sustainable supply chain operations prioritize the optimization of packaging materials to reduce waste. This involves the utilization of eco-friendly packaging and the minimization of unnecessary packaging (Zhu & Sarkis, 2007).
- ***Circular economy practices:*** The circular economy (CE) presents a fresh and distinct perspective on organizational and operational systems of production and consumption. It centers on restoring the value of used resources, offering economic, environmental, and social benefits. The circular economy (CE) is posited as a novel business mindset that can facilitate the transition of organizations and society toward sustainable development. Countries such as China, several European nations, and Japan have instituted incentive regulations to motivate organizations to adopt CE principles. The incorporation of circular economy principles in supply chain management promotes the recycling and reuse of products and materials (Geissdoerfer et al., 2017).
- ***Recycled materials procurement:*** Sustainable supply chains prioritize the procurement of products made from recycled materials, thereby stimulating market demand for recycling and encouraging the utilization of eco-friendly products (Sarkis & Dhavale, 2015).

Implementing waste reduction and recycling practices in SSCM, as presented in prior practices outlined in this chapter, can be a challenging process. It necessitates aligning waste management goals with sustainability objectives and investing in recycling infrastructure and technologies, typically entailing a significant financial investment.

6.6 The Evolution of Sustainable Operation Management

Based on the literature review we consider that five approaches are most important in the evolution of Sustainable Operation Management:

6.6.1 The Reverse Supply Chain

Originally, SCM was perceived as a process-oriented and customer-focused business discipline in which material flows are directed from upstream suppliers to downstream customers. More recently, the concepts of reverse logistics and closed-loop SCM have complemented the traditional perception of a "forward SC" through product-recycling aspects and the recovery of scarce materials (Brandenburg et al., 2019). For organizations to incorporate sustainability considerations need to make strategic changes that are related to the reverse supply chain. The reverse supply chain aims to collect and reuse end-of-life products returned by customers, structuring supply chain incentives to properly motivate partners, and managing competition between remanufactured and new products (Gupta & Palsule-Desai, 2011). Examining the literature on the reverse supply chain, it is evident that addressing the challenge of selecting an appropriate reverse channel structure for collecting used products involves three options: (i) direct collection from customers, (ii) incentivizing existing retailers for collection, or (iii) subcontracting the collection to a third party. This analysis demonstrates that straightforward coordination mechanisms can achieve the same level of collection effort by the retailer and supply chain profits as in a centrally coordinated system (Gupta & Palsule-Desai, 2011). Yet, a "return SC" is not necessarily a sustainable one, thus, the socio-ecological dimensions need to be emphasized in the explanation of sustainable supply chain management (SSCM), which is comprehended as "the management of material, information, and capital flows, as well as cooperation among companies along the supply chain, while taking goals from all three dimensions of sustainable development, i.e., economic, environmental, and social, into account, which are derived from customer and stakeholder requirements" (Brandenburg et al., 2019). However, in the literature, there are some critiques toward the use of the reverse model in the supply chain, which means that they should use the downstream and upstream flows of goods (Brandenburg & Rebs, 2015).

6.6.2 Closed-Loop Supply Chain Management

Closed-loop SCM (CLSCM) describes the interplay of various factors formally in a mathematical way but in contrast, SSCM research is mainly elaborated qualitatively based on conceptual frameworks elaborated by Carter and Rogers (2008) and empirically by case studies and surveys (Brandenburg & Rebs, 2015). The closed-loop supply chain model emphasizes the reduction, reuse, and recycling of resources. It explores how businesses can design supply chains that minimize waste, energy consumption, and environmental impact.

6.6.3 Triple-Bottom-Line (TBL) Model

The TBL model, also known as the 3Ps (People, Planet, Profit), is a foundational framework for SOM. It explores the three dimensions of sustainability and how organizations can balance economic success with social responsibility and environmental stewardship. The triple-bottom-line (TBL) approach involves addressing conflicts or capitalizing on the synergy among economic, environmental, and social factors. The economic dimension encompasses both financial and non-financial goals, while socio-environmental objectives seek to enhance positive ecological impacts and address social aspects. Firms often encounter challenges in pursuing sustainability transformation, requiring them to navigate trade-offs among different TBL dimensions. A win–win perspective, while beneficial, may limit strategic options tied to immediate profit logic, hindering radical shifts in business practices (Brandenburg et al., 2019).

6.6.4 Green Supply Chain Management

Green Supply Chain Management focuses on integrating sustainability into the entire supply chain. Green supply chain management (GSCM) integrates eco-friendly practices seamlessly into traditional supply chains, spanning from manufacturing and operations to end-of-life management. It operates on the principle of 4R1D, emphasizing the reduction, reuse, recycling, reclamation, and degradation of resources.

6.6.5 Lean and Green Operations

Integrating Lean principles with green practices aims to minimize waste, optimize resource utilization, and reduce environmental impact. A Lean and Green Production System is characterized as a system that concurrently and continuously improves operational efficiency and sustainability. Its primary objective is to significantly enhance the organization or supply chain's capability to generate and deliver value to customers and society at large. Several principles underpin the integration of Lean and Sustainability Principles:

– ***Efficient Resource Use:*** Lean practices inherently promote efficient resource utilization, as waste elimination minimizes unnecessary resource consumption and diminishes the environmental footprint (Sarkis & Dhavale, 2015).

- ***Reduction of Carbon Emissions:*** Incorporation of Lean supply chains usually results in optimized transportation and more efficient inventory management which leads to carbon emissions reduction.
- ***Minimizing Waste:*** Rooted in the Toyota Production System, lean principles concentrate on eradicating waste in all forms, encompassing overproduction, excess inventory, or inefficient processes (Womack & Jones, 1996). These principles aim to enhance customer value delivery while minimizing resource utilization.
- ***Sustainable Procurement:*** The integration of lean and sustainability principles extends to green procurement, entailing the sourcing of products and services that meet lean efficiency requirements while adhering to eco-friendly and ethical standards (Womack & Jones, 1996).

6.7 Challenges in Sustainable Operations and Supply Chain Management

Among the foremost challenges in sustainable operations and supply chain management (SOSCM), a prominent concern is the absence of standardized metrics and practices. The lack of universally accepted sustainability standards impedes benchmarking, making it challenging to assess and compare the sustainability performance of diverse organizations (Seuring & Müller, 2008). The establishment of consensus on standardized metrics becomes imperative for fostering transparency and accountability in the realm of sustainable operations and supply chain management. Additional challenges in SOSCM encompass:

- ***Global Complexity:*** In the era of globalization, organizations grapple with the intricacy of managing sustainable practices across diverse global supply networks. Varied regulatory environments, cultural nuances, and differing levels of sustainability awareness pose challenges in ensuring consistent and uniform sustainability practices throughout the supply chain (Carter & Easton, 2011). Navigating this complexity necessitates a nuanced understanding of global dynamics and the formulation of tailored strategies for sustainable operations.
- ***Cost Considerations:*** The implementation of sustainable practices may entail initial costs, potentially acting as a barrier for some businesses.
- ***Supply Chain Complexity:*** Managing sustainability across the entire supply chain, particularly for businesses with complex global networks, presents challenges. Ensuring alignment with sustainability goals across all suppliers requires vigilant oversight and effective communication.
- ***Changing Regulations:*** Environmental and labor regulations exhibit wide variation across regions and evolve over time. Businesses must remain informed and adaptive to the dynamic nature of legal requirements.
- ***Resistance to Change:*** Resistance to change within organizational structures stands as a formidable challenge in adopting sustainable practices. Employees, suppliers, and other stakeholders may resist shifts toward sustainability due to

perceived disruptions, cost implications, or lack of awareness (Zhu et al., 2008). Overcoming this resistance necessitates effective change management strategies and the cultivation of a culture valuing sustainability.

- ***Advanced Technologies:*** The integration of advanced technologies, including blockchain, artificial intelligence, and the Internet of Things, is pivotal for enhancing transparency and traceability in sustainable operations and supply chains (Mollenkopf et al., 2010). However, the adoption of these technologies poses challenges in terms of investment costs, data security, and the need for organizational readiness.

Understanding and addressing these challenges are crucial for organizations aspiring to establish more resilient and responsible supply chains. The future of sustainable operations lies in collaborative efforts, technological innovations, and a collective commitment to advancing the principles of sustainability.

6.8 Benefits of Sustainable Operations and Supply Chain Management

Organizations often make decisions regarding the adoption of new approaches, methodologies, and technologies based on anticipated benefits. Despite the increasing implementation of sustainability practices by many countries through legislative measures, articulating and promoting the positive outcomes derived from incorporating sustainable operations and supply chain management (SOSCM), as evidenced by previous research, expedites the acceptance of proposed approaches. There are many benefits from implementing the mentioned programs, some of the more preeminent benefits encompass:

a) **Cost Savings:** A foremost and immediately tangible benefit of SOSCM is cost savings. Much research proves that these investments frequently result in prolonged cost savings and advantages. In the realm of sustainable operations, the design of facility layouts has been recognized as a crucial factor influencing operational performance, particularly in manufacturing systems (Tayal et al., 2017).
b) **Risk Mitigation:** Sustainability initiatives play a pivotal role in helping organizations mitigate various risks, including supply chain disruptions, regulatory penalties, and reputational damage. Through supplier diversification and the adoption of sustainable practices, companies enhance their resilience to unforeseen challenges (Sarkis, 2016).
c) **Customer Loyalty and Market Access:** Ethically minded consumers are inclined to remain loyal to brands aligning with their values. Additionally, certain markets and procurement opportunities are exclusively accessible to companies boasting strong sustainability credentials (Sarkis & Dhavale, 2015).
d) **Compliance with Regulations:** Sustainable operations frequently align with existing and emerging environmental and social regulations. Proactive compli-

ance with these regulations enables organizations to sidestep penalties and legal entanglements (Sarkis & Dhavale, 2015).

e) **Resource Efficiency and Long-Term Viability:** SOSCM prompts organizations to utilize resources efficiently, prolong product life cycles, and reduce waste. This not only supports environmental objectives but also ensures the long-term viability of the business by conserving resources (Sarkis, 2016).

Sustainable operations and supply chain management offer numerous advantages to organizations willing to integrate sustainability as a foundational element of their business strategies, ranging from financial savings to enhanced reputation.

6.9 Summary

The chapter provides a comprehensive academic-level overview of the evolving landscape of sustainability in business and industry. It underscores the transformation of sustainability from a mere corporate responsibility to a strategic imperative, crucial for addressing environmental, social, and economic challenges. Emphasis is placed on sustainable supply chain management (SSCM), dissecting its three pillars: Environmental sustainability, social responsibility, and economic efficiency. The discourse acknowledges research gaps, particularly inconclusive findings on the correlation between SSCM practices and organizational performance, attributing it to the reliance on managerial perceptions rather than concrete performance data.

Strategies for SSCM are outlined, ranging from green procurement to lean and efficient processes, energy efficiency and renewable energy, and waste reduction and recycling. The narrative recognizes the challenges associated with these strategies, including cost considerations, resistance to change, and the need for cultural transformation. Despite the hurdles, the text asserts the imperative for organizations to adopt sustainability practices, citing environmental, social, and economic factors as driving forces for long-term success.

The challenges in sustainable operations and supply chain management (SOSCM) are dissected, encompassing the absence of standardized metrics, global complexity, cost considerations, supply chain complexity, changing regulations, resistance to change, and the integration of advanced technologies. Addressing these challenges is deemed crucial for organizations aspiring to establish resilient and responsible supply chains.

The benefits of SOSCM are systematically delineated, encompassing cost savings, risk mitigation, reputation enhancement, innovation, competitive advantage, customer loyalty, market access, compliance with regulations, resource efficiency, long-term viability, stakeholder engagement, and employee satisfaction. The text advocates for a holistic understanding of these benefits while acknowledging the need for mitigating potential challenges.

Case Study: The True Price of Roses from Kenya

A comprehensive analysis was conducted to determine the viability of sustainable rose farming through a true price assessment. The investigation focused on T-hybrid roses sourced from Lake Naivasha in Kenya, comparing conventional farming practices with sustainable ones. By examining the entire supply chain, it was revealed that the retail prices per stem for roses from both farm types averaged €0.70. However, the true price, accounting for environmental and social factors, was significantly lower for sustainable roses (€0.74) compared to conventional roses (€0.92). The primary contributors to this difference were the environmental impact of airfreight transportation and the social impact on workers' incomes.

The true price analysis identified several projects to mitigate environmental and social costs, including shipping by sea to reduce carbon emissions, implementing solar-powered greenhouses, adopting closed-loop hydroponics for water and fertilizer efficiency, providing health and safety training for workers, establishing gender committees to address harassment and discrimination, and ensuring a basic living wage for improved worker well-being.

Cost mapping of each project was performed, revealing their impact on the profit and loss of an average farm. For instance, health and safety training could generate approximately €4500 profit per hectare, while transitioning to sea transport could increase profit by €5000 per hectare. Importantly, implementing better social standards and environmentally friendly practices proved financially feasible without adversely affecting farm owners' bottom lines.

However, some improvements in social standards, such as paying a living wage, were deemed less feasible if borne solely by farm owners. An economic value chain analysis demonstrated that providing a living wage could be achievable when sharing the costs among wholesale traders, retail traders, and consumers. This approach reinforced the advancement of superior social and environmental standards within the industry.

Questions:

1. What were the key findings of the true price analysis in the sustainable rose farming case study, and how did the true price compare between sustainable and conventional roses?
2. What were the primary contributors to the difference in true price between sustainable and conventional roses, as identified in the case study?
3. How did the case study propose to mitigate environmental and social costs in sustainable rose farming, and what were the projected impacts of specific projects on the profit and loss of an average farm?
4. In terms of social standards, particularly regarding the provision of a living wage, what challenges were identified in the case study, and how did the economic value chain analysis suggest addressing these challenges through shared costs among wholesale traders, retail traders, and consumers?

Source: Authors. Adopted from Schoenmaker (2017).

Questions for Discussion

- Define sustainable operation management and explain its main goal!
- Define the sustainable chain management!
- Explain the three pillars of sustainable supply chain management and point to differences between them!
- Explain the strategies for sustainable operations and supply chain management and how they can help family businesses implement sustainability programs related to operation management!
- Explain the main approaches toward the evolution of sustainable operation management!
- Name and explain the main challenges in sustainable operations and supply chain management!

Additional Readings

- Schoenmaker, D., & Schramade, W. (2018). *Principles of sustainable finance*. Oxford University Press.
- Lehner, O. M. (Ed.). (2016). *Routledge handbook of social and sustainable finance*. Routledge.
- Rexhepi, G. (2016). The architecture of social finance. *Routledge handbook of social and sustainable finance*, 35–49.

Suggested Activities

- Divide the class into small groups and assign each group one of the challenges discussed in the chapter, such as the lack of standardized metrics, global complexity, or resistance to change. The team should present their findings to the class, fostering a comprehensive discussion on the multifaceted challenges associated with sustainable operations and supply chain management.
- Formulate a hypothetical scenario where a company is facing sustainability challenges in its operations and supply chain management, divide the class into teams, with each team representing the management of the hypothetical company, and instruct teams to develop a comprehensive sustainability strategy that addresses the challenges identified in the chapter.
- Organize a debate where students are divided into two groups: one advocating for the continued emphasis on sustainable operations and supply chain management, and the other exploring potential criticisms or alternative viewpoints.
- Debate among students with different specific roles to each student within the groups, such as industry executives, environmentalists, consumers, or policymakers, to ensure a diverse range of perspectives.

Keywords and Concepts

- Sustainable operations
- Sustainable chain management
- Corporate responsibility
- Sustainable business models
- Environmental sustainability

- Resource efficiency
- Emission reduction
- Circular economy principles
- Social responsibility
- Ethical labor practices
- UN Guiding Principles on Business and Human Rights
- Economic efficiency
- Lean manufacturing
- Green procurement
- Eco-friendly suppliers
- Ethical labor practices
- Energy efficiency
- Lean and efficient processes
- Renewable energy
- Waste reduction
- Circular economy practices
- Supply chain complexity
- Risk mitigation
- Reputation enhancement
- Compliance with regulations

References

Bazilian, M., et al. (2011). Re-considering the economics of photovoltaic power. *Renewable Energy, 36*(1), 145–150.

Bocken, N. M., Short, S. W., Rana, P., & Evans, S. (2014). A literature and practice review to develop sustainable business model archetypes. *Journal of Cleaner Production, 65*, 42–56.

Brandenburg, M., Gruchmann, T., & Oelze, N. (2019). Sustainable supply chain management—A conceptual framework and future research perspectives. *Sustainability, 11*(24), 7239.

Brandenburg, M., & Rebs, T. (2015). Sustainable supply chain management: A modeling perspective. *Annals of Operations Research, 229*, 213–252.

Braungart, M., & McDonough, W. (2002). *Cradle to cradle: Remaking the way we make things*. North Point Press.

Carter, C. R., & Easton, P. L. (2011). Sustainable supply chain management: Evolution and future directions. *International Journal of Physical Distribution & Logistics Management, 41*(1), 46–62.

Carter, C. R., & Rogers, D. S. (2008). A framework of sustainable supply chain management: Moving toward new theory. *International Journal of Physical Distribution & Logistics Management, 38*(5), 360–387.

Chen, J., Hu, X., Razi, U., & Rexhepi, G. (2022). The sustainable potential of efficient air-transportation industry and green innovation in realising environmental sustainability in G7 countries. *Economic research-Ekonomska istraživanja, 35*(1), 3814–3835.

Cooper, M. C., Lambert, D. M., & Pagh, J. D. (1997). Supply chain management: More than a new name for logistics. *The international journal of logistics management, 8*(1), 1–14.

Dora, M., Wesana, J., Gellynck, X., Seth, N., Dey, B., & De Steur, H. (2020). Importance of sustainable operations in food loss: Evidence from the Belgian food processing industry. *Annals of Operations Research, 290*, 47–72.

Elkington, J. (1998). *Cannibals with forks: The triple bottom line of 21st century business*. Capstone.

Epstein, M. J. (2008). *Making sustainability work: Best practices in managing and measuring corporate social, environmental, and economic impacts*. Berrett-Koehler Publishers.

Freeman, R. E. (1984). *Strategic management: A stakeholder approach*. Pitman.

Geissdoerfer, M., et al. (2017). The circular economy—A new sustainability paradigm? *Journal of Cleaner Production, 143*, 757–768.

Goli, A., & Mohammadi, H. (2022). Developing a sustainable operational management system using hybrid Shapley value and Multimoora method: Case study petrochemical supply chain. *Environment, Development and Sustainability, 24*(9), 10540–10569.

Govindan, K. (2018). Green supply chain management: A state-of-the-art literature review. *International Journal of Production Research, 56*(1–2), 299–316.

Graafland, J., et al. (2014). Motives of socially responsible business conduct. *Journal of Business Ethics, 121*(3), 425–441.

Green, K., Zelbst, P., Meacham, J., & Bhadauria, V. S. (2012). Green supply chain management practices: Impact on performance. *Supply Chain Management: An International Journal, 17*(3), 290–305.

Gunasekaran, A., & Irani, Z. (2014). Sustainable operations management: Design, modelling and analysis. *Journal of the Operational Research Society, 65*, 801–805.

Gupta, S., & Palsule-Desai, O. D. (2011). Sustainable supply chain management: Review and research opportunities. *IIMB Management Review, 23*(4), 234–245.

Kajtazi, K., Rexhepi, G., Sharif, A., & Ozturk, I. (2023). Business model innovation and its impact on corporate sustainability. *Journal of Business Research, 166*, 114082

Lai, K. H., & Wong, C. W. (2012). Green logistics management and performance: Some empirical evidence from Chinese manufacturing exporters. *Omega, 40*(3), 267–282.

Mollenkopf, D., Stolze, H., Tate, W. L., & Ueltschy, M. (2010). Green, lean, and global supply chains. *International Journal of Logistics Management, 21*(3), 360–383.

Murphy, P. R., & Poist, R. F. (2012). Green purchasing and supply policies: Do they improve companies' environmental performance? *International Journal of Production Research, 50*(11), 2986–3000.

Myers, K. & Conte, N., 2013. Can social finance improve the outcomes of employment and training programs? /, Canadian Electronic Library. Ottawa, Ontario. Retrieved from https://canadacommons.ca/artifacts/1193845/can-social-finance-improve-the-outcomes-of-employment-and-training-programs/1746969/ on 08 Jun 2024. CID: 20.500.12592/1s1z6m.

Rahdari, A., Sheehy, B., Khan, H. Z., Braendle, U., Rexhepi, G., & Sepasi, S. (2020). Exploring global retailers' corporate social responsibility performance. *Heliyon, 6*(8).

Rexhepi, G., Abazi-Alili, H., Abduli, S., Ibraimi, S., & Zuferi, R. (2023). Green entrepreneurship and firm performance: the case of Albania. In Entrepreneurship development in the balkans: perspective from diverse contexts (pp. 69–80). Emerald Publishing Limited.

Sarkis, J. (2012). A boundaries and flows perspective of green supply chain management. *Supply Chain Management: An International Journal, 17*(2), 202–216.

Sarkis, J. (2016). Corporate environmental sustainability and DEA. In *Handbook of Operations Analytics Using Data Envelopment Analysis* (pp. 483–498).

Sarkis, J., & Dhavale, D. G. (2015). Supplier selection for sustainable operations: A triple-bottom-line approach using a Bayesian framework. *International Journal of Production Economics, 166*, 177–191.

Schoenmaker, D. (2017). *Investing for the common good: A sustainable finance framework* (p. 80). Bruegel.

Schultmann, F., & Sun, M. (2019). Green supply chain management: Past, present, and future trends. *International Journal of Production Research, 57*(3), 829–846.

Sepasi, S., Rexhepi, G., & Rahdari, A. (2021). The changing prospects of corporate social responsibility in the decade of action: Do personal values matter?. *Corporate Social Responsibility and Environmental Management, 28*(1), 138–152.

Seuring, S., & Gold, S. (2013). Sustainability management beyond corporate boundaries: From stakeholders to performance. *Journal of Cleaner Production, 56*, 1–6.

Seuring, S., & Müller, M. (2008). From a literature review to a conceptual framework for sustainable supply chain management. *Journal of Cleaner Production, 16*(15), 1699–1710.

Srivastava, S. K. (2007). Green supply-chain management: A state-of-the-art literature review. *International Journal of Management Reviews, 9*(1), 53–80.

Suki, N. M., Suki, N. M., Sharif, A., Afshan, S., & Rexhepi, G. (2023). Importance of green innovation for business sustainability: Identifying the key role of green intellectual capital and green SCM. *Business Strategy and the Environment, 32*(4), 1542–1558.

Taticchi, P., Tonelli, F., & Cagnazzo, L. (2013). Performance measurement and management: A literature review and a research agenda. *Measuring Business Excellence, 17*(1), 4–18.

Tayal, A., Gunasekaran, A., Singh, S. P., Dubey, R., & Papadopoulos, T. (2017). Formulating and solving sustainable stochastic dynamic facility layout problem: A key to sustainable operations. *Annals of Operations Research, 253*, 621–655.

Walker, H., & Jones, N. (2012). Sustainable supply chain management across the UK private sector. *Supply Chain Management: An International Journal, 17*(1), 15–28.

Womack, J. P., & Jones, D. T. (1996). *Lean thinking: Banish waste and create wealth in your corporation*. Simon and Schuster.

Yazan, D. M., & Fraccascia, L. (2020). Sustainable operations of industrial symbiosis: An enterprise input-output model integrated by agent-based simulation. *International Journal of Production Research, 58*(2), 392–414.

Zhu, Q., & Sarkis, J. (2007). The moderating effects of institutional pressures on emergent green supply chain practices and performance. *International Journal of Production Research, 45*(18–19), 4333–4355.

Zhu, Q., Sarkis, J., & Lai, K. H. (2007). Green supply chain management: Pressures, practices and performance within the Chinese automobile industry. *Journal of Cleaner Production, 15*(11–12), 1041–1052.

Zhu, Q., Sarkis, J., & Lai, K. H. (2008). Confirmation of a measurement model for green supply chain management practices implementation. *International Journal of Production Economics, 111*(2), 261–273.

R&D Innovation in Family Businesses

7

Learning Objectives
After reading this chapter, you should be able to:

- Define R&D in family firms.
- Define the 3Ps in family firms.
- Understand the role of R&D in family firms in regard to societal impact.
- Understand the role of R&D in family firms in regard to economic impact.
- Understand the role of R&D in family firms in regard to environmental impact.
- Explain the implication of R&D in family firms regarding the 3Ps.

Profile: LG "Sciencepark" for Global R&D[1]

LG Corporation, a prominent South Korean business, achieved the 18th position in the Family Business Index of 2023, as reported by the Center for Family Business of the University of St. Gallen. The company traces its roots back to its establishment in Seoul where it originated as Lucky Chemical Co., Ltd. The initial focus was on cosmetics, marked by the establishment of the first cosmetics laboratory. This early foundation has since evolved, positioning LG Corporation as a notable player in various industries.

In 1951, LG Corporation kicked off a new era in the chemical and electronics industries. Fast forward to October 1958, and Goldstar Co., Ltd was born, which we now know as LG Electronics. The big moment came in 1959 when Goldstar produced the first-ever Korean radio, proudly named A-501. By

[1] Written by Ensar Mekic, Sultan Qaboos University.

R. Palalić et al., *Sustainable Family Business*, Springer Texts in Business and Economics, https://doi.org/10.1007/978-3-031-57999-8_7

1960, LG was all about establishing key industries, and as the years rolled on, LG kept adding new innovative products to its lineup such are: the first 6-transistor radios in Korea, followed by 4-tube radios, and then 12-inch fans, a first in Korea. In addition, the company introduced the first automatic telephone in Korea. These moves show LG's capabilities for breaking new ground and bringing cutting-edge tech to the Korean market.

From 1970 to 1994, LG witnessed a high-growth period driven by cutting-edge technology development and robust internal management. Goldstar Co., Ltd. achieved several noteworthy milestones during this time, including the development and production of the first Korean cassette recorder, 19-inch color TVs, the first electronic VCR in Korea, the first domestic minicomputers in Korea, the first color video camera in Korea, and the first projection TV in Korea. In 1984, a pivotal moment occurred as Bando Corp. underwent a name change to become Lucky Goldstar International Corp. The transition into the twenty-first century, spanning from 1995 to 2017, marked a significant move toward "Superior LG."

In 1995, the name of Lucky Goldstar Group was simplified to LG. The development continued in the coming two decades. In 2015, LG Electronics made a remarkable move by introducing so-called "LG Signature," a line of super-premium household appliances, in Korea. This year also saw a milestone for LG Electronics, with washer sales reaching an impressive 150 million units. Meanwhile, LG Chem played a part in space exploration by supplying lithium-ion batteries for NASA space suits. Additionally, their "Gram 15" notebook computer secured a spot in the Guinness Book as the world's lightest notebook.

In 2022, LG Electronics introduced the world's largest OLED TV, measuring an impressive 97 inches. Moving into 2023, the company took a proactive stance on environmental responsibility by releasing the "Net Zero Special Reporting" along with a Carbon Neutral Action Plan. Notably, LG's OLED TV secured its position as the undisputed leader in the global OLED TV market for the remarkable milestone of 10 consecutive years. Emphasizing a commitment to sustainability, LG presented the "LG Smart Cottage" at IFA 2023, advocating for an eco-friendlier lifestyle. Further, in an exciting tech leap, LG Electronics launched Korea's first foldable laptop, the "LG Gram Fold." These initiatives underscore LG's dedication to innovation, environmental consciousness, and pushing boundaries in the technology landscape.

Considering so many successful innovation stories, one may ask what is LG's recipe for research and development, especially if we consider its complexity due to having a network of affiliated companies in electronics (LG Electronics, LG Display, LG Innotek), in chemicals (LG Chem, LG Energy Solution, LG Household and Health Care), and communication and services (LG U+, LG Hello Vision Corp., HS Ad, LG CNS, D&O, LG Management Development Institute, LG Sports, LG Corp).

The key role for LG's R&D is played by LG Sciencepark through a community of innovators pioneering new futures for LG and its customers. Since its establishment in 2018, LG Sciencepark has evolved into the premier research and development hub within the LG Group. Spanning over 1 million square meters (equivalent to 152 football pitches), it stands as Korea's largest R&D complex. More than just a physical space, LG Sciencepark is home to a collaborative community of over 20,000 individuals, representing eight LG affiliate companies, partner firms, and startups. The LG Sciencepark fosters an open community that encourages exploration and the realization of entirely new possibilities for a better life. According to LG, in today's dynamic landscape, an open and collaborative approach to innovation is not just a preference but a necessity. Foremost in its mission is the facilitation of collaboration among LG affiliate companies, aiming to foster innovative synergy. Across electronics, chemicals, communication, and services, the organization is committed to unlocking the vast potential within its diverse business and R&D portfolio.

Finally, LG pursues an ESG strategy, underscoring its commitment to responsible business practices for a sustainable future. The company asserts its dedication to delivering positive impacts across economic, environmental, social, and governance aspects. This commitment is manifested through initiatives focused on climate action, circular economy practices, responsible supply chain management, safety and health standards, and the promotion of fair business practices. In embracing these principles, LG aims to contribute meaningfully to a more sustainable and inclusive global landscape.

Bibliography

Center for Family Business. (2023). 2023 Family Business Index. (University of St.Gallen) Retrieved December 2023, from https://familybusinessindex.com

LG. (2023). www.lgcorp.com. Retrieved 12 2023, from https://www.lgcorp.com/about/companies

LG. (2023). www.lgcorp.com. Retrieved 12 2023, from https://www.lgcorp.com/about/history/1

LG. (2023). https://www.lgcorp.com. Retrieved 12 2023, from https://www.lgcorp.com

LG. (2023). www.lgcorp.com. Retrieved 12 2023, from https://www.lgcorp.com/innovation/sciencepark/introduce

LG. (2023). www.lgcorp.com. Retrieved 12 2023, from https://www.lgcorp.com/innovation/sciencepark/mission

LG. (2023). www.lgcorp.com. Retrieved 12 2023, from https://www.lgcorp.com/esg/strategy

7.1 Introduction

Research and development (R&D) is one of the supporting activities (Rothaermel, 2017) (like IT, HR, and accounting and finance) within a venture that develops and produces a new product in the market. As Rothaermel (2017) discussed that R&D and innovation are connected but in terms of costs, they may differ in its structure and amount. Logically, investment in R&D seems to boost innovation and the R&D investment will be that much reflected in new and innovative products. Innovation has nothing to do with how many R&D dollars you have. When Apple came up with the Mac, IBM was spending at least 100 times more on R&D. It is not about money. It is about the people you have, how you are led, and how much you get it (p. 145). This thoughtful example can be intrigued and milestone for the family business owners.

As discussed in Chap. 9, leadership plays an important role in integrating business strategies through the prism of the TBL approach, so is vital for R&D strategy formulation. As family businesses rely always on people, who create value for the family and the businesses, the family firm leaders should take the role of R&D seriously and perhaps follow the examples of innovative firms like Apple, where people matter. High intensive industries in which family firms compete, the R&D investment becomes a necessity (Block, 2012).

The following sections address issues and challenges posed to family-owned firms in implementing TBL through their R&D and how they eventually balance the TBL qualities.

7.1.1 R&D and Challenges for Family Firms

To gain a competitive advantage and sustain it in the long term, family firms compete through innovative products, which are based on innovations. As regular SMEs family firms tend to have various partnerships at different levels and in different places, regions, and countries. Innovative products offered by family firms are often driven either by customers demand, unserved markets, or a societal problem in the community. Keeping it up-to-date is the real challenge for family firms due to strong competition in the industry, in which innovation plays a significant role. As non-family firms are open to external collaboration in which they complement their current know-how, knowledge, competence, or expertise, that much the family firms may see that as a challenge. A challenge that can prevent family businesses from being open to R&D stimuli is socioemotional wealth. As SEW keeps the family business firm within the family tradition and values (Astrachan & Jaskiewicz, 2008), the non-financial (Zellweger & Astrachan, 2008), it means being open to external resources of social capital that can foster innovation within the family firm.

7.1.2 R&D Agenda to Implement TBL Approach

The UN agenda of 2030 (UN, 2023), challenged all firms to respond to this call of being responsible for their own planet. With these 17 sustainable development goals (SDGs), family businesses are challenged by this call from one side, and they have been advantages by this call from another. The challenge may imply formulating a new strategy that will quest a new corporate culture to comprehensively respond to this call because it seems that family ownership is not fully ready (Gharbi &

Othmani, 2021) to positively respond to this call. The advantage, on the other hand, is because family businesses were being a constituent part of the community and its symbiotic life (Dana et al., 2000) with the environment.

Therefore, integrating R&D into TBL approach implies aligning all R&D activities accordingly, embracing the 3Ps (planet, people, profit) responsibilities toward environment, society, and economic perspective.

7.2 Nature of R&D in Family Businesses

Research and development (R&D) in family firms often depicts unique characteristics determined by the unique characteristics (Distelberg & Blow, 2010; Ramadani & Hoy, 2015) of these firms. Unlike non-family firms, family businesses have inherited sense of being deeply involved in entrepreneurial activities, which are at the same time reflected in their R&D activities.

As family firm owners are regarded as entrepreneurs, R&D in their firms is not an unknown notion. Therefore, implementing the R&D strategy that will be in compliance to planet, people, and profit (3Ps) will not be that challenging, but rather a turning point for founders to take the lead in this call to save the planet, people, and earn profit. In other words, the game rule has changed recently, which family businesses should follow and act accordingly.

7.3 R&D AND Societal Impact: Family Business Perspective

Family brand image in the community needs to stay positive (Zellweger et al., 2012), strong, and meaningful, as for the people, as well as for employees (family members employees and non-family members employees). Evidence from 2006 shows that family firms are more socially responsible than non-family firms (Dyer & Whetten, 2006). Recently, Ramadani and his associates elaborated on this matter providing evidence that family businesses are embedded in the society and they try to be one body with the closed community (Ramadani et al., 2020). Having said that, family business needs to incorporate in their strategy the 3Ps. They need to establish (Fig. 7.1):

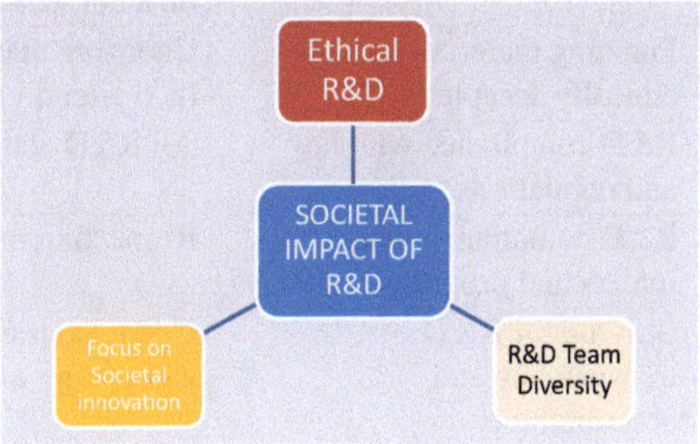

Fig. 7.1 R&D and societal impact: a family business perspective. Source: Authors (based on the text above)

- **An ethical research and development**
- **A diverse team in R&D**
- **A focus on societal innovation**

7.3.1 Establishing Ethical Research and Development

Corporate strategy should involve a strategy that will formulate **code of ethics** in the research and development. It means that the code of ethics should encompass all stakeholders in the R&D process, like labor, sourcing suppliers (materials), compliance with laws and regulations, paying attention to intellectual property, being transparent and accountable, as well as other social norms and practices rooted in ethically accepted activities (Table 7.1).

Ethical labor practices involve ensuring that all employees and contractors involved in the R&D process are treated on a fair and ethical basis, including fair payments (wages and salaries), safe working conditions, and reasonable working hours. Additionally, a family firm's strategy should avoid favoritism and be professional in which employment and professional promotion will be based on merits.

Pursuing materials that are ethically accepted, by requiring sourcing materials in a way that is socially and environmentally safe. In other words, a family firm should choose supplier(s) who adhere to ethical practices in sustainability of sources of materials supply, and the chain in general. Moreover, family firms should avoid anything that will harm environment, people, or infringe on any rights in these activities.

R&D compliance with Laws and Regulations, family firms ensure that the R&D activities are legally and ethically sound. It includes compliance with labor laws and regulations as well as industry standards.

R&D compliance with intellectual property (IP) includes respecting the intellectual property rights of others. Meaning, the R&D process avoids plagiarism-related issues by respecting patents and copyrights, or usage of third-party research are properly credited and licensed.

Table 7.1 Ethical R&D in family firms. Source: Authors

Name	Description
Ethical labor practices	All stakeholders (employees, partners, suppliers) are treated on a fair and ethical basis
Pursuing materials that are ethically accepted	Choosing materials that will be environmentally safe (eco-friendly)
R&D compliance with law and regulations	All R&D activities should be ethically sound
R&D compliance with intellectual property	Respecting intellectual property of third parties
Striving for R&D transparency and accountability	Open and transparent R&D in regard to methodology and technology used in developing innovative products, etc.

Striving for R&D transparency and accountability involves openly sharing research methodologies, results, and potential impacts of the products or technologies developed, which is essential while respecting accountability for the social and environmental impacts of R&D activities. Furthermore, the R&D practices in family firms should consider other social impacts like assessing how new products or technologies will affect users, communities, and society at large, especially in terms of privacy, security, health, and well-being.

To succeed in their strategies, family businesses must ensure that the corporate culture encourages and promotes ethical behavior at all levels by making clear and transparent policies and procedures in the above-mentioned areas. This is where family firms' leadership needs to uphold these standards and to make decisions that will preserve the family tradition, values, but also reflect positively on broader societal norms.

7.3.2 Diversity in Research and Development Team

Diversity in R&D team is of crucial importance. It can be mandatory inclusion within the family firm because the family firm can create a good knowledge base on its own. In other words, it can create its own social capital (Arregle et al., 2007). Arregle and associates argue that two types of social capital exist in family firms. These are the ones from *family* and the other one is from the *firm*. Arregle et al. (2007) define social capital based on the argument of Adler and Kwon (2002) which is a tight relationship between individuals and organizations in which they create value (p. 75).

For the family's social capital, they hold that in a family, members exchange support, information, and influence, fostering loyalty and unity. This environment of mutual support and shared resources, including monetary and economic interests, strengthens the family's social bond. Such strong family bonds, seen as a form of *social capital*, are truly enduring and influential in helping family members succeed.

In contrast to families, new business organizations typically lack strong stability and interconnected relationships. Despite this, they require more social capital than families to manage complex challenges and uncertainties. Therefore, the success of new businesses depends significantly on building and maintaining robust social networks. A joint co-existence of the social capital made by family and non-family members brings better social capital than the one made of family members only (Sanchez-Famoso et al., 2019). In this case, R&D gets its full potential in terms of diversity (Fig. 7.2).

7.3.3 Focus on Societal Innovation

Societal innovation should be focused on community's needs that have a positive impact on its socio-economic development. In other words, such community-focused innovations in family businesses require the development of products or

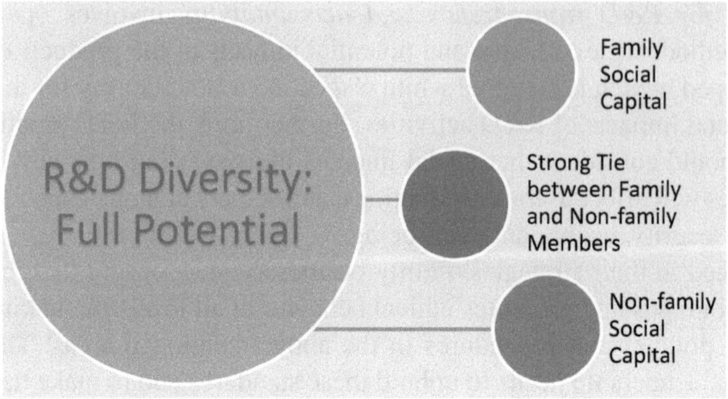

Fig. 7.2 Diversity in R&D in family firms: Full potential. Source: Authors (based on the text above)

services that directly address the specific needs of the communities served. This approach is based on a deep understanding of local issues, creating links between business and its social context. By focusing on locally relevant problems and issues, such innovations can make a significant contribution to society's well-being.

Moreover, it could be providing affordable products for low-income people in the community, developing local environmental solutions, or providing services that improve the health and education of locals. Additionally, such an immense community engagement in the innovation process provides family businesses valuable reputation that may have impact in the long term. Hence, these reciprocal or symbiotic relationships (Dana et al. 2000) reinforce the business's commitment to social well-being and develop a loyal customer base based on shared community values.

7.4 R&D and Economic Impact: Family Business Perspective

The new world agenda argues that firms can earn profit while paying attention to the society and environment. In economic terms, firms strive to get sustainable competitive advantage through innovations along all processes and products. In order to achieve that, all firms respect the fact that competitive advantage takes along the risk, costs, and market needs.

Family firms' R&D, like in any other type of firms, should have an economic impact while addressing other two postulates of the TBL approach. Therefore, R&D activities should be in accordance to:

- Market needs
- Cost leadership
  - Cost management
  - Cost mitigation

7.4.1 Market Needs and R&D

R&D investment in family-owned businesses is contingent upon various constraints. A primary factor is the "agency problem," as outlined by Chi (2023), Choi et al. (2015), and Morck and Yeung (2003). This issue arises from the family's natural inclination toward avoiding risks and financial limitations, which discourages investments in R&D. When this problem exists, there is the tendency for a reduction in R&D spending to maintain the family's wealth and control, by excessive control (Chi, 2023), leading family firms to adopt a more traditional and cautious approach to business operations. "Wait and see posture" is embraced (Barringer & Ireland, 2019).

Investment in R&D in family firms depends on many factors. The first one is the "agency problem" (Chi, 2023; Choi et al., 2015; Morck & Yeung, 2003), which can prevent family firms from investing in R&D activities because families have risk-aversion and some kind of wealth constraints that prevent such investments. So if the agency problem is strong, less will be invested in R&D due to preserving wealth and "excessive control right," which makes the family firm a kind of "conservative" private entity.

Nonetheless, when considering the broader perspective of R&D's benefits and its alignment with the Triple-Bottom-Line (TBL) approach—which emphasizes environmental, social, and economic sustainability—family firms must anticipate market demands. These demands often highlight unaddressed societal needs (problems). By addressing these needs, family firms not only adhere to TBL principles but also capitalize on market opportunities, thus delivering both societal value and financial returns. This strategic approach to R&D ensures that consumer needs are met and market gaps are fulfilled.

This focus on the market is critical to growing the business and maintaining its competitive edge. Specifically, investment in R&D involves striking a balance between innovation, societal norms, and market realities to ensure that R&D efforts lead to commercial breakthroughs.

7.4.2 Cost Management and Cost Mitigation in R&D

Cost management is crucial in all types of firms and organizations (Yoon et al., 2021). While non-family firms are more open to external stakeholders and investors, family businesses are less interested in exploiting external opportunities to enhance innovation. Costs must be contained to some extent but family firms might have a myopic view of investment in R&D, regarding it as the major cost.

Due to socioemotional wealth (SEW) preservation (Kalm & Gomez Mejia, 2016), family firms tend to have less investment in R&D than non-family firms, unless economic goals and family goals converge (Chrisman & Patel, 2012). In such cases, the cost are being asymmetric (increase less while sales increase more) with

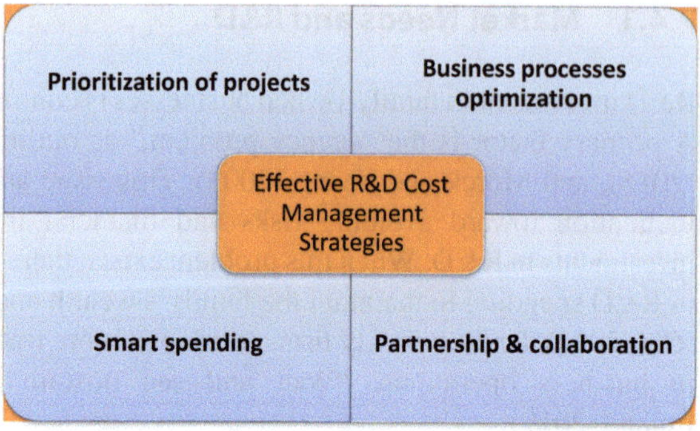

Fig. 7.3 Strategies for effective cost management for R&D in family firms. Source: Authors

sales volume, which represents the "asymmetric cost behavior" (Yoon et al., 2021), or "cost stickiness" (Anderson et al., 2007). **Cost stickiness or asymmetric cost behavior** is defined when costs tend to go up more easily when business activity increases, but do not decrease as much when business activity drops by the same amount.

The bottom line of cost management in R&D in family firms is a critical aspect for businesses, particularly in maintaining a balance between fostering innovative approaches and adhering to budget constraints. Being effective cost management, it involves *several key strategies* (Fig. 7.3).

First, **prioritization of projects**. As not all R&D projects will yield equal value, it is extremely important to allocate resources to projects with the highest potential for success and return on investment (ROI). This prioritization usually involves a thorough feasibility analysis, as well as risk forecast to identify projects that align with business goals and have a viable market and financial potential.

Second, **business process optimization**. This encompasses the lean methodology (Klein et al., 2022; Pombal et al., 2019) to prioritize and optimize the whole value chain processes starting from eliminating waste, optimizing workflows, and efficient and effective resource allocation. Besides reducing cost of production, for example, it also accelerates time schedule of prototype which enables a quicker market entry and getting a return on the created value from the market.

Third, **smart spending**. This implies a smart investment (mitigating unnecessary costs) in areas with the highest potential impact. This could include investing in high-quality materials or advanced technologies that will have a positive impact in the long term. In other words, investing more at the start may be more expensive, but it often results in greater success and profits over time. It is important to remember that these costs are really long-term investments, not just immediate expenses. Family ownership may be negative toward R&D investments (Choi et al., 2015; Delgado-García et al., 2023), but it may be positive regarding the smart investment (Chen & Hsu, 2009)

Fourth, **partnerships and collaborations**. Family firms can use other leverages to reduce costs by collaborating with external bodies like universities, research institutes, or other companies and organizations, which will share costs and risks of R&D. Any collaborative ties (Dong et al., 2021) can lead to R&D progress to enhance the innovative outcomes in projects that are the most important (prior ones)

Family firms are kin to contain the cost, and thus they need to have a control function that will evaluate the investment (spending) in R&D and the outcome of that investment, the ROI.

7.5 R&D and Its Impact on Environment: Family Business Perspective

Several years ago, Berrone and his associates (Berrone et al., 2010) found that family-controlled public firms in the United States strengthen their socioemotional wealth (SEW) via better performance in environmental activities, regardless of whether the CEO is a founder or he plays both roles, the founder and CEO. This is one of the strong arguments for family firms that they intend to preserve their image in society (Ramadani et al., 2020; Babin et al., 2017). They are aware that it is important and that society sees them through environmental and societal lenses. One of the ways family firms create wealth is to embed their family name in society through different environmental and societal activities, which consequently strengthens their SEW (Lumpkin & Bacq, 2022). Moreover, this shows how family firms are deeply involved in "civic wealth creation"(CWC) (Lumpkin & Bacq, 2022). *Civil wealth creation* is "a comprehensive indicator of the intellectual, affective, and material resources, capacities, and capabilities of a civic unit of analysis" (Lumpkin & Bacq, 2019: 386).

Therefore, such a family-owned firm's role in society reflects in strengthening the following three very important wealth: *socioemotional wealth, civic wealth, and financial wealth. Socioemotional wealth* is a non-financial factor that influences the business performance of family firms and it affects the overall commitment of the family members in the firm (Berrone, et al., 2012). It is "non-financial aspects of the firm that meet the family's affective needs" (Gómez-Mejía et al., 2007; Gómez-Mejía & Herrero, 2022). Financial wealth is an accumulation of all financial values created by the firm throughout the whole life cycle.

In order to sustain the above types of wealth, family businesses need to establish their corporate strategy that will emphasize the following:

- **Innovation in R&D that will be reflected in eco-friendly products**
- **Optimization of the current resource allocation**
- **Regular assessment of R&D performance (monitoring)**

7.5.1 Innovation in R&D: Eco-friendly Products

Long-term strategies of family firms should be in line with eco-friendly innovation, which means their R&D should focus on developing environmentally friendly products and processes, such as renewable energy technologies, biodegradable materials, and waste reduction techniques (Jafri & Liu, 2023).

Renewable energy technologies quest R&D to focus on exploiting energy from renewable natural sources like sun, wind, and water (Jafri & Liu, 2023). These technologies are critical to reducing dependence on fossil fuels (Zhang et al., 2018) and reducing carbon emissions while addressing global warming and climate change. Therefore, constant availability of renewable resources for sustainable fuel production is essential for family firms.

A great example is Singapore, setting the Green Plan 2030 (Green Plan 2030, 2023).[2] The "Green Plan, is a whole-of-nation movement to advance Singapore's national agenda on sustainable development." This "Green Plan charts ambitious and concrete targets over the next 10 years, strengthening Singapore's commitments under the UN's 2030 Sustainable Development Agenda and Paris Agreement,[3] and positioning us to achieve our long-term net zero emissions aspiration by 2050." This agenda calls for businesses to respond to the national plan to use the resources efficiently in their business activities, specifically in R&D activities to save the planet and build better well-being of people.

7.5.2 Optimized Allocation of the Current Resources

European Commission (EC) formulated a long-term strategy to reduce carbon emissions by 2050 (European Commission's Green Deal, 2019).[4] The European Community (EC) is focusing on resource efficiency, aiming to enhance the use of resources in research and development (R&D) effectively. This means less waste and a smaller ecological footprint. The trend in competitiveness is shifting toward eco-friendly products and processes that use sustainable technologies, materials, and energy. Family businesses, in particular, are well-positioned to achieve this goal due to their long-standing commitment to honorable practices in their community and regions.

7.5.3 Regular Monitoring and Assessment of R&D Activities

Family businesses usually manage all their operations closely to ensure they are done well and without waste. These firms often prefer to avoid risks or only take on risks that they can manage. Therefore, they need to keep a regular check on their R&D work and how it is performing. They should also continuously check how their R&D affects the environment and try to make this impact better over time. This kind of continuous assessment aims to create products that are not harmful to the environment and help tackle today's environmental problems and challenges.

[2] Singapore green plan. Available at: https://www.greenplan.gov.sg/ Accessed on December 15, 2023.

[3] Paris Agreement: Available at: https://unfccc.int/process-and-meetings/the-paris-agreement. Accessed on December 15, 2023. The Paris Agreement is a legally binding international treaty on climate change. It was adopted by 196 Parties at the UN Climate Change Conference (COP21) in Paris, France, on December 12, 2015. It entered into force on November 04, 2016.

[4] European Comission, 2019: Available at: https://climate.ec.europa.eu/eu-action/climate-strategies-targets/2050-long-term-strategy_en#eu-strategy Accessed on December 15, 2023.

7.6 Implications of R&D Within the Triple-Bottom Approach in Family Businesses

R&D in family businesses is characterized by a long-term, value-based approach, influenced by family dynamics and operating constraints. This results in unique innovation that measures equality in balancing traditional values with the need to modernize and adapt to changing market demands.

Implications of the R&D in family businesses could be seen throughout the framework, planet, people, and profit (3Ps):

1. R&D and planet (environmental aspect)
2. R&D and people (societal aspect)
3. R&D and profit (economic aspect)

Environmental aspect of R&D. The environmental aspect of R&D in family businesses focuses on creating new products that are not only innovative but also environmentally friendly. These businesses aim to address environmental problems and meet market demands by incorporating eco-friendly practices into their product development strategies.

Societal aspect of R&D. In creating innovative and environmentally friendly products, family-owned businesses aim to address social challenges and contribute positively to society. They leverage their strong community reputation to fulfill their commitment to societal well-being while preparing to tackle potential future global problems that could impact local people.

Economic aspect of R&D. The economic aspect is just as crucial as the environmental and social sides for family businesses. The profit from selling innovative and eco-friendly products is essential. To succeed economically, these businesses should focus on what the market needs and practice cost leadership. This means reducing unnecessary expenses and managing costs effectively, especially in their research and development operations. Family businesses need to integrate these economic strategies into their overall company plans. This creates a balance between meeting the global demand for sustainable practices (TBL approach) while preserving its history, tradition, values, and positive image in society.

R&D at Bayerische Motoren Werke AG (BMW)[5]

Bayerische Motoren Werke AG (BMW) earned the 6th spot in the 2021 ranking based on Family Business Index (Center for Family Business, 2021). BMW produces cars and motorcycles, including famous brands like BMW, MINI, Rolls-Royce, and BMW Motorrad. What sets BMW apart is its strong focus on being eco-friendly and saving resources. They make sure to be

[5] Written by Ensar Mekic, Sultan Qaboos University.

careful about where they get their materials and how they make their products. This commitment to sustainability goes from the very start of producing products to the end of their life cycle.

BMW from 1916 to Electric Vehicles Era

The BMW Group has a really long history, going back to 1916 and filled with inventors, pioneers, and engineers. BMW's success with motorcycles began in 1923 at the Berlin Motor Show when they introduced their first motorcycle, the BMW R 32. People loved it because it was a well-made machine. In 1928, BMW decided to make cars too. They bought a company called Fahrzeugfabrik Eisenach, and their first car after World War II was the 501, starting in 1952. Later in 1972, BMW created a new subsidiary called BMW Motorsport GmbH to bring together all their motor racing activities. In 1982, they decided to build a new plant in Regensburg because more people wanted the BMW 3 Series, and they needed extra space. BMW has kept getting better, and now they are a big player in the electric vehicle market. Since they launched the BMW i3 in 2013, they have been adding more and more electric vehicles to their lineup.

Today, BMW proudly stands as a global leader in crafting premium cars and motorcycles, complemented by esteemed financial and mobility services. With an extensive presence across over 30 production sites worldwide and a widespread sales network, BMW strives to provide excellence. In 2022, its customers had the privilege of choosing from an impressive array of eleven fully electric variants spanning eight distinctive model ranges. This diverse electric portfolio marks a significant presence in the premium segment. What distinguishes BMW in the realm of electrification is the company's commitment to developing and producing electric drive systems in-house, ensuring a unique and high-quality experience.

Performance Results in 2022

In 2022, the BMW Group achieved remarkable results, selling nearly 2.4 million cars and over 202,000 motorcycles globally. The pre-tax profits for the year reached €23.5 billion, with revenues totaling €142.6 billion. The company's global workforce is composed of 149,475 dedicated individuals as of December 31, 2022.

Beyond its financial success, BMW prioritizes its responsibility to the environment, society, and its employees. The company is guided by a commitment that goes beyond profit, aiming to make a positive impact. Oliver Zipse, Chairman of the Board of Management of BMW AG, underlines this commitment, stating, *"Sustainability and economic success go hand in hand at the BMW Group. As a premium manufacturer, we have the ambition to lead the way in the area of sustainability."*

Research & Development at BMW

The significance of BMW's research and development efforts has always been paramount. In 1986, BMW AG consolidated all its research and development activities into the Research and Innovation Centre in Munich. Pioneering the automotive industry, it was the first of its kind, bringing together approximately 7000 professionals, including scientists, engineers, designers, managers, and technicians. This integrated team collaborated seamlessly under one roof, marking a groundbreaking approach to innovation. The official inauguration of this facility took place on April 27, 1990.

For over a century, the BMW Group has been synonymous with pioneering innovation, woven into the fabric of its past, present, and future. In essence, BMW is reshaping the driving experience for the digital era. The company's rich history reveals a tradition of remarkable innovations, with research and development playing a pivotal role in this legacy. This forward-looking philosophy has not only defined but also propelled the company to numerous milestones. Below are some examples of the most noteworthy results that showcase BMW's commitment to pushing the boundaries of innovation:

- 1917: BMW introduces the BMW IIIa, featuring the first-ever aircraft engine with aluminum pistons.
- 1935: At the German Motor Show, BMW unveils the BMW R12 and BMW R17, showcasing the world's first hydraulic front fork for motorcycles.
- 1936: BMW produces the BMW 328, hailed as the most successful sports car of the late 1930s.
- 1954: BMW pioneers the type 502, presenting the world's premier aluminum V8 engine for series production.
- 1957: BMW establishes a dedicated market research division, demonstrating a commitment to understanding customer needs.
- 1973: The BMW head office building designed with a unique high-rise suspension supported by four pillars earns recognition as a monument.
- 1973: BMW pioneers environmental responsibility by appointing the first environmental officer in the automotive sector.
- 1983: BMW introduces first-ever diesel car, setting new standards with an exceptional six-cylinder engine.
- 1988: A groundbreaking collaboration with FAG and Hella leads to the launch of the first electronically controlled antilock braking system for series-produced motorcycles.
- 1991: BMW achieves a significant environmental milestone by introducing the world's first series-produced three-way catalytic converter in the BMW K1.
- 1991: Paving the way for electric vehicles, BMW unveils the first fully-fledged thoroughbred electric car of the modern age.
- 2001: BMW takes the lead in innovation by becoming the first manufacturer worldwide to introduce a fully-variable valve lift system in the new and innovative VALVETRONIC engine in the BMW 316i.

The BMW Group's Technology Offices are strategically positioned in key technology and innovation hubs worldwide. These locations, often situated in major global tech centers, allow the company to stay at the forefront of emerging technologies. BMW's customers globally reap the rewards of having on-site experts who delve into cutting-edge technologies well before they hit the market. The potential of these technologies is thoroughly explored through pilot projects, and the insights gained are then shared across various central divisions. Importantly, the scope of innovation extends beyond vehicle development. The Technology Offices are instrumental in driving advancements across diverse business areas, ranging from production processes to developing solutions for smart cities and shaping the future of mobility.

The expansive network of Technology Offices globally holds a pivotal role in maintaining the BMW Group's position as an innovation leader and is a cornerstone of our open innovation strategy. For the BMW Group, open innovation embodies:

- *Actively Identifying Trends:* Proactively seeking out and embracing new technology trends through initiatives like the Technology Trend Radar.
- *Leveraging Innovation Potential:* Tapping into both external and internal sources of innovation, including startups, intrapreneurs, cross-industry technologies, and collaborations with universities.
- *Expanding Beyond Core Business:* Embracing a mindset that goes beyond the core automotive business, extending innovation efforts to realms beyond traditional car-related domains.
- *Building Collaborations:* Establishing and fostering new collaborations, exemplified by initiatives such as the BMW Startup Garage, to cultivate partnerships that drive innovation.
- *Agile Action:* Operating swiftly and with agility, implementing proof of concept projects and developing minimum viable products to test and iterate on innovative ideas efficiently.
- *Sustainable Transfer of Innovations:* Ensuring the sustainable integration of innovations into the BMW Group, facilitating a seamless transition from concept to implementation.

The overarching goals of BMW Group Technology Offices can be succinctly summarized as follows: (1) Innovation Scouting and Trend Reports whereby the company actively scouts and identifies relevant innovations, generating comprehensive trend reports to stay at the forefront of emerging technologies. (2) Strategic Networking which enables BMW to establish meaningful connections with startups, universities, and cross-industry partners, engaging in proof-of-concept initiatives and research projects to foster collaborative innovation. (3) Effective Knowledge Transfer to facilitate the seamless transfer of pertinent innovations to the central organization, ensuring that valuable insights and advancements are integrated into the broader framework of the BMW Group.

In the current digital age of mobility, the automotive industry grapples with a sophisticated approach to car manufacturing, treating vehicles as high-tech products known as intelligent cars. These cars not only deliver exceptional performance but also offer extensive connectivity features. A pivotal advancement in this domain is the adoption of the new 5G mobile radio standard, which significantly enhances data transmission capabilities compared to its predecessor. Pioneering this technological frontier, the BMW Group achieved a milestone with the BMW iX. As the first premium manufacturer globally, BMW successfully integrated the 5G mobile radio standard into a production vehicle, underlining its commitment to staying at the forefront of technological innovation in the automotive landscape.

One of the crucial components within BMW's organizational structure is the Open Innovation unit. Kai Petric, who leads this unit, expressed: *"At BMW Open Innovation we create value for our customers by unlocking the innovation potential of startups, cross industry technologies, intrapreneurs and innovation crowds. Thanks to our highly committed team and a comprehensive outside-in perspective we significantly contribute to the innovation leadership of the BMW Group."*

The BMW Startup Garage stands as another pivotal unit within the BMW Group—functioning as the venture client unit. Its primary mission revolves around identifying innovations that offer tangible advantages for the BMW Group's products, services, systems, and processes. This initiative is designed to assess startups and establish them as enduring partners for the BMW Group, contributing significantly to the company's ongoing innovation leadership and reinforcing its commitment to staying at the forefront of industry advancements.

The BMW Crowd Innovation Platform serves as a dynamic space where innovators can harness the collective intelligence of both internal and external innovation crowds. This collaborative approach allows for the precise identification of new ideas and technical solutions, providing a proactive means of addressing current challenges and anticipating solutions for future problems.

The technology scouts at the BMW Group operate on an international stage, collaborating both internally with technology offices in the United States, Asia, and Europe, and externally with a diverse range of partners such as trend agencies, consultancies, and higher education institutions. Their overarching objective is the early identification of novel and innovative technologies, as well as technology-driven trends. By doing so, they aim to make these advancements applicable to the BMW Group and gain a comprehensive understanding of customer demands for future mobility solutions in global markets.

In our contemporary world, cars are transforming into drivers of high-level digital experiences, data collection, and delivering outstanding performance with optimal efficiency. Observing how this German family business tackles these challenges, it becomes evident that their approach to Research and Development (R&D) results in excellent outcomes that are not only customer-friendly but also environmentally conscious. The innovative strides made by this family business showcase a commitment to creating products that not only meet the evolving needs of consumers but also align with sustainability and environmental considerations.

Discussion Questions:

- What are the key innovation enabling mechanisms at BMW?
- Considering the industry, is BMW's sustainability-oriented R&D a choice, or a must?
- Is BMW's agenda in line with the TBL principles? Why or why not?
- How BMW's R&D approach to innovation benefits people, planet and its profits?

Bibliography

BMW. (n.d.). *www.bmwgroup.com.* Retrieved 12 21, 2023, from https://www.bmwgroup.com/en/company/history.html

BMW. (n.d.). *www.bmwgroup.com.* Retrieved 12 21, 2023, from https://www.bmwgroup.com/en/company/brands-products.html

BMW. (n.d.). *www.bmwgroup.com.* Retrieved 12 21, 2023, from https://www.bmwgroup.com/en/company.html

BMW. (n.d.). *www.bmwgroup.com.* Retrieved 12 21, 2023, from https://www.bmwgroup.com/en/sustainability.html

BMW. (n.d.). *www.bmwgroup.com.* Retrieved 12 21, 2023, from https://www.bmwgroup.com/en/innovation.html

BMW. (n.d.). *www.bmwgroup.com.* Retrieved 12 21, 2023, from https://www.bmwgroup.com/en/innovation/open-innovation.html

BMW. (n.d.). *www.bmwgroup.com.* Retrieved 12 21, 2023, from https://www.bmwgroup.com/en/innovation/open-innovation/startup-garage.html

BMW. (n.d.). *www.bmwgroup.com.* Retrieved 12 21, 2023, from https://www.bmwgroup.com/en/innovation/open-innovation/crowd_innovation.html

BMW. (n.d.). *www.bmwgroup.com.* Retrieved 12 21, 2023, from https://www.bmwgroup.com/en/innovation/open-innovation/technology-scouting.html

Center for Family Business. (2021, June). Family businesses across the world. Switzerland: Center for Family Business at the University of St. Gallen. Retrieved from https://2021.familybusinessindex.com/#intro

Questions for Discussion

- Explain the nature of R&D in family firms.
- Discuss the societal aspect of R&D in family firms.
- Explain and give appropriate examples of effective cost management strategies of R&D in family firms.
- Discuss the economic aspect of R&D in family firms. Give appropriate examples.
- Discuss the environmental aspect of R&D in family firms. Give appropriate examples.

Additional Readings

- Dong, F., Wang, X., & Chen, J. (2021). Family ownership and cooperative R&D: the moderating effect of political ties. Journal of Knowledge Management, 26(2), 403-422. https://doi.org/10.1108/jkm-08-2020-0651
- Berrone, P., Cruz, C., Gomez-Mejia, L. R., & Larraza-Kintana, M. (2010). Socioemotional Wealth and Corporate Responses to Institutional Pressures: Do Family-Controlled Firms Pollute Less? Administrative Science Quarterly, 55(1), 82-113. https://doi.org/10.2189/asqu.2010.55.1.82
- Ramadani, V. & Hoy, F. (2015), Context and uniqueness of family businesses. In Dana, L-P. & Ramadani, V. (Eds.), Family Businesses in Transition Economies: Management, Succession and Internationalization (pp 9-37), Cham: Springer.

Suggested Activities

- Identify one or more family-owned firms and compare them with non-family firms in regard to R&D. What do you see as the main difference? Explain it with appropriate examples.
- Find one family and one non-family businesses. Compare their R&D performance and investments. Explain differences and reasons for their symmetry or their asymmetry.

Summary of Key Concepts and Keywords

Agency problem

- Asymmetric cost behavior
- Cost leadership strategy
- Cost stickiness
- Effective cost management strategy
- Ethical R&D
- Market needs
- Research and development (R&D)
- R&D diversity
- Smart spending
- Societal innovation

Websites

United Nations, 2023: https://sdgs.un.org/goals Accessed on 12 December 2023.

European Commission, 2019: Available at: https://climate.ec.europa.eu/eu-action/climate-strategies-targets/2050-long-term-strategy_en#eu-strategy Accessed on 15 December 2023.

Singapore green plan. Available at: https://www.greenplan.gov.sg/ Accessed on 15 December 2023.

The Family Business Consulting Group. Available at: https://www.thefbcg.com/Secrecy%2D%2DA-Family-Business-Vice/. Accessed on 21.02.2020.

References

Adler, P. S., & Kwon, S. W. (2002). Social capital: Prospects for a new concept. *Academy of Management Review, 27*, 17–40.

Anderson, M., Banker, R., Huang, R., & Janakiraman, S. (2007). Cost behavior and fundamental analysis of SG&A costs. *Journal of Accounting, Auditing and Finance, 22*(1), 1–28. https://doi.org/10.1177/0148558X0702200103

Arregle, J. L., Hitt, M. A., Sirmon, D. G., & Very, P. (2007). The development of organizational social capital: Attributes of family firms*. *Journal of Management Studies, 44*(1), 73–95. https://doi.org/10.1111/j.1467-6486.2007.00665.x

Astrachan, J. H., & Jaskiewicz, P. (2008). Emotional returns and emotional costs in privately held family businesses: Advancing traditional business valuation. *Family Business Review, 21*(2), 139–149.

Babin, B. J., Binz Astrachan, C., Botero, I. C., Hair, J. F., Prügl, R., Herrmann, J.-L., Laroche, M., & Sarstedt, M. (2017). From family identity to family firm image and reputation: Exploring facets of the perception of family influence in branding, marketing, and other messaging. *Journal of Family Business Strategy, 8*(2), I–III. https://doi.org/10.1016/s1877-8585(17)30125-0

Barringer, B. R., & Ireland, R. D. (2019). *Entrepreneurship: successfully launching new ventures* (6, Global edition ed.). Pearson.

Berrone, P., Cruz, C., Gomez-Mejia, L. R., & Larraza-Kintana, M. (2010). Socioemotional wealth and corporate responses to institutional pressures: Do family-controlled firms pollute less? *Administrative Science Quarterly, 55*(1), 82–113. https://doi.org/10.2189/asqu.2010.55.1.82

Berrone, P., Cruz, C., & Gomez-Mejia, L. R. (2012). Socioemotional wealth in family firms: Theoretical dimensions, assessment approaches, and agenda for future research. *Family Business Review, 25*(3), 258–279. https://doi.org/10.1177/0894486511435355

Block, J. H. (2012). R&D investments in family and founder firms: An agency perspective. *Journal of Business Venturing, 27*(2), 248–265. https://doi.org/10.1016/j.jbusvent.2010.09.003

Chen, H.-L., & Hsu, W.-T. (2009). Family ownership, board independence, and R&D investment. *Family Business Review, 22*(4), 347–362. https://doi.org/10.1177/0894486509341062

Chi, Y.-L. (2023). The agency costs of family ownership: Evidence from innovation performance. *Journal of Banking & Finance, 148*. https://doi.org/10.1016/j.jbankfin.2022.106737

Choi, Y. R., Zahra, S. A., Yoshikawa, T., & Han, B. H. (2015). Family ownership and R&D investment: The role of growth opportunities and business group membership. *Journal of Business Research, 68*(5), 1053–1061. https://doi.org/10.1016/j.jbusres.2014.10.007

Chrisman, J. J., & Patel, P. C. (2012). Variations in R&D investments of family and nonfamily firms: Behavioral agency and myopic loss aversion perspectives. *Academy of Management Journal, 55*(4), 976–997. https://doi.org/10.5465/amj.2011.0211

Dana, L. P., Etemad, H., & Wright, R. W. (2000). The global reach of symbiotic networks. *Journal of Euro Marketing, 9*(2), 1–16. https://doi.org/10.1300/J037v09n02_01

Delgado-García, J. B., Blanco-Mazagatos, V., Romero-Merino, M. E., & Díaz-Portugal, C. (2023). Family CEO affect and R&D investments of family firms: The moderation effect of family ownership structure. *Long Range Planning, 56*(5). https://doi.org/10.1016/j.lrp.2022.102230

Distelberg, B., & Blow, A. (2010). The role of values and unity in family businesses. *Journal of Family and Economic Issues, 31*, 427–441. https://doi.org/10.1007/s10834-010-9221-4

Dong, F., Wang, X., & Chen, J. (2021). Family ownership and cooperative R&D: the moderating effect of political ties. *Journal of Knowledge Management, 26*(2), 403–422. https://doi.org/10.1108/jkm-08-2020-0651

Dyer, W. G., & Whetten, D. A. (2006). Family firms and social responsibility: Preliminary evidence from the S&P 500. *Entrepreneurship Theory and Practice, 30*, 785–802.

Gharbi, S., & Othmani, H. (2021). Family ownership and R&D investment: the moderating role of institutional investors. *International Journal of Corporate Governance, 12*(2), 185–207.

Gómez-Mejía, L. R., & Herrero, I. (2022). Back to square one: The measurement of Socioemotional Wealth (SEW). *Journal of Family Business Strategy, 13*(4). https://doi.org/10.1016/j.jfbs.2021.100480

Gómez-Mejía, L. R., Haynes, K. T., Núñez-Nickel, M., Jacobson, K. J. L., & Moyano-Fuentes, J. (2007). Socioemotional wealth and business risks in family-controlled firms: Evidence from Spanish olive oil mills. *Administrative Science Quarterly, 52*(1), 106–137. https://doi.org/10.2189/asqu.52.1.106

Jafri, M. A. H., & Liu, H. (2023). Eco-innovation and its influence on renewable energy demand: The role of environmental law. *International Journal of Environmental Research and Public Health, 20*(4). https://doi.org/10.3390/ijerph20043194

Kalm, M., & Gomez Mejia, L. R. (2016). Socioemotional wealth preservation in family firms. *Revista de Administração, 51*, 409–411.

Klein, L. L., Vieira, K. M., Feltrin, T. S., Pissutti, M., & Ercolani, L. D. (2022). The influence of lean management practices on process effectiveness: A quantitative study in a public institution. *Sage Open, 12*(1), 21582440221088837. https://doi.org/10.1177/21582440221088837

Lumpkin, G. T., & Bacq, S. (2019). Civic wealth creation: A new view of stakeholder engagement and societal impact. *The Academy of Management Perspectives, 33*(4), 383–404.

Lumpkin, G. T., & Bacq, S. (2022). Family business, community embeddedness, and civic wealth creation. *Journal of Family Business Strategy, 13*(2). https://doi.org/10.1016/j.jfbs.2021.100469

Morck, R., & Yeung, B. (2003). Agency problems in large family business groups. *Entrepreneurship Theory and Practice, 27*(4), 367–382. https://doi.org/10.1111/1540-8520.t01-1-00015

Pombal, T., Ferreira, L. P., Sá, J. C., Pereira, M. T., & Silva, F. J. G. (2019). Implementation of lean methodologies in the management of consumable materials in the maintenance workshops of an industrial company. *Procedia Manufacturing, 38*, 975–982. https://doi.org/10.1016/j.promfg.2020.01.181

Ramadani, V., & Hoy, F. (2015). Context and uniqueness of family businesses. In L. P. Dana & V. Ramadani (Eds.), *Family businesses in transition economies*. Springer.

Ramadani, V., Memili, E., Chang, E. P., & Palalić, R. (2020). *Entrepreneurial family businesses: Innovation, governance and succession* (1st ed.). Springer.

Rothaermel, F. (2017). *Strategic management* (3rd ed.). Mc Graw Hill Education.

Sanchez-Famoso, V., Pittino, D., Chirico, F., Maseda, A., & Iturralde, T. (2019). Social capital and innovation in family firms: The moderating roles of family control and generational involvement. *Scandinavian Journal of Management, 35*(3). https://doi.org/10.1016/j.scaman.2019.02.002

Yoon, S., Chung, Y., & Son, H. (2021). The relationship between cost behaviour and R&D efficiency: Empirical evidence from Korean stock markets. *Technology Analysis and Strategic Management, 34*(1), 85–98. https://doi.org/10.1080/09537325.2021.1887474

Zellweger, T. M., & Astrachan, J. H. (2008). On the emotional value of owning a firm. *Family Business Review, 21*(4), 347–363. https://doi.org/10.1177/08944865080210040106

Zellweger, T. M., Kellermanns, F. W., Eddleston, K. A., & Memili, E. (2012). Building a family firm image: How family firms capitalize on their family ties. *Journal of Family Business Strategy, 3*, 239–250.
Zhang, X., Lovati, M., Vigna, I., Widén, J., Han, M., Gal, C., & Feng, T. (2018). A review of urban energy systems at building cluster level incorporating renewable-energy-source (RES) envelope solutions. *Applied Energy, 230*, 1034–1056.

Global Operations in Family Businesses

8

Learning Objectives
After reading this chapter, you should be able to:

- Define the term globalization.
- Define global operations.
- Address the importance of sustainability of family firms in global operations via TBL.
- Understand tools and mechanisms of global operations of family firms using TBL.

Profile: Dell Technologies Inc.[1]
Dell Technologies Inc., a global leader in technology, media, and communications, attained the ninth position in the 2023 Family Business Index, according to the Center for Family Business at the University of St. Gallen. Recognized as one of the world's premier technology companies, Dell Technologies is dedicated to transforming lives through extraordinary capabilities.

Their impact is far-reaching, encompassing innovative solutions such as hybrid cloud technologies, high-performance computing, and substantial contributions to social impact and sustainability initiatives, influencing people worldwide. Dell Technologies' development timeline starts with 1984, and more details on company's continuous improvement until today have been presented below:

[1] Ensar Mekic, Sultan Qaboos University. Based on the bibliography used.

R. Palalić et al., *Sustainable Family Business*, Springer Texts in Business and Economics, https://doi.org/10.1007/978-3-031-57999-8_8

- 1984—Michael Dell established PC's Limited with a mere $1000 and an innovative vision.
- 1987—Dell opens its first international subsidiary in the United Kingdom.
- 1988—Dell opens a new integrated sales, manufacturing, and support center in Xiamen, China.
- 1988—Dell opens a new manufacturing operations center in Brazil.
- 2001—Dell becomes the No. 1 computer systems provider worldwide and reaches No. 1 in US Intel-based server shipments.
- 2004—Dell became China's third largest provider of computer systems and services.
- 2013—In 2013, Michael Dell and private equity firm Silver Lake Partners buy back Dell from public shareholders.
- 2023—Company enjoys a reputation as a leading company in manufacturing laptops, desktops, workstations, monitors, servers, storage, and electronics.
- 2023—Dell Technologies is instrumental in changing the digital landscape the world over.

The company's current revenue in 2023 stands at $91.14 billion. In the fiscal year 2022, Dell experienced a total revenue of $105.25 billion, reflecting an increase compared to the preceding year's revenue of $105.11 billion in 2021. The remarkable financial accomplishments of Dell owe much to its expansive global business operations. Operating one of the largest end-to-end supply chains worldwide, the company effectively manages a network of over 24,000 suppliers. Dell ships an impressive 75 million units annually, with support extended through more than 2180 global service centers and collaboration with over 28 global manufacturing locations. In other words, the company dispatches two units every second, translating to a staggering 179,000 orders daily. The global operations of the company provide its employees with a truly international experience. As articulated by Diane Z., a senior manager at Dell, "I've worked on three continents, led multiple projects, and influenced a Fortune 500 supply chain." This testimonial emphasizes the transformative impact of Dell's global operations on the professional journeys of its personnel.

Finally, it is important to emphasize that Dell cares for sustainability and consider it as necessity in global operations. In 2022, Dell demonstrated its commitment to sustainability by incorporating 39 million kilograms (86 million pounds) of sustainable materials into its products and packaging. Fueled by innovation, the company's overarching mission is to protect and enhance the planet, improve conditions for those involved in the production of its products, and minimize its environmental impact. Impressively, 94.5% of all packaging materials now consist of recycled or renewable materials. Dell's dedication to environmental responsibility extends beyond packaging. Since 2019, the company has successfully prevented 39.2 million plastic bottles from entering the ocean, based on the standard half-liter bottle. This achievement reflects Dell's commitment to mitigating plastic pollution and its adverse effects on marine ecosystems. Moreover, Dell proudly sources energy responsibly. Currently, 59% of the electricity used in its facilities comes from renewable sources, showcasing a commitment to cleaner and more sustainable energy practices. These achievements collectively underscore Dell's unwavering dedication to sustainability, emphasizing its ongoing efforts to make a positive impact on the environment and the global community.

Bibliography

Center for Family Business. (2023). 2023 Family Business Index. (University of St.Gallen) Retrieved December 2023, from https://familybusinessindex.com

Dell. (2023). Dell Technologies Global Operations. Retrieved December 2023, from www.dell.com: https://www.dell.com/en-us/lp/dell-tech-global-ops

Dell. (2023). Who we are. Retrieved December 2023, from www.dell.com: https://www.dell.com/en-ph/dt/corporate/about-us/who-we-are.htm?hve=discover+our+brands

Dell. (2023). Dell Technologies Timeline. Retrieved December 2023, from www.dell.com: https://www.dell.com/en-ph/dt/corporate/about-us/who-we-are/timeline.htm?hve=explore+our+history

Companies Marketcap. (2023). Revenue for Dell (DELL). Retrieved December 2023, from www.companiesmarketcap.com: https://companiesmarketcap.com/dell/revenue/

8.1 Introduction

The term globalization was first coined by Theodore Levitt, an American economist with German roots, in 1983 in his famous article "The Globalization of Markets." Recent work by Volle (2023)[2] argues that globalization, in modern terms, started immediately after the industrial revolution in nineteenth century. However, some authors claim that the globalization took place many years before. Going back to ancient history, we can observe that movement of people, goods, and money (which is nowadays defined as globalization) had happened many centuries back. One of the facts is that the "silk road" existed centuries ago, which linked China with Europe (via Turkiye, so called "Anatolia"), India, North Africa, and Middle East.[3]

[2] Volle, A. (2023, November 29). globalization. Encyclopedia Britannica. https://www.britannica.com/money/topic/globalization

[3] Encyclopedia Britannica. https://www.britannica.com/money/topic/Silk-Road-trade-route. Accessed on December 16, 2023.

Today's' globalization started in mid of twentieth century, which was enhanced by the newest technology and communication that eased its progress till now. Thanks to this event, movement of goods and services was not faster than now. This opened the door for businesses to operate across borders instead of competing only in the local market. This brought advantages for local firms because they could compete somewhere else and exploit other benefits, but also it brought, on the other hand, competition to local firms making the growth difficult.

Nonetheless, the newest technology and communication enhanced rapid firms' growth and development. These developments in technology and communication make it easier for products and services to move from one region to another, while all these activities were followed by marketing activities that paved the way for success in foreign markets.

While shifting to foreign operations, firms developed strategies that will respond to the specific region a firm enters in. This is called customizations in which firms adjust or adapt their products or services to the context in which they operate. To succeed, they customize their products and services according to the culture that dominates its needs, wants, and demands. However, as global operations got a bigger scale, which quested an enormous usage of natural and other resources that has its consequences, the time came to revisit the way the businesses are done. Greediness for money has alerted the world to respond to create a better future for the whole planet.

A recent new world agenda led by the United Nations (UN) which introduced the sustainable development goals (SDGs),[4] calls the whole globe (organizations, institutions, and individuals) to adjust their business operations and activities in a more balanced manner. While being better off in economic terms (profit), firms need to respect the environment in which people live (planet), as well as the society (people), known as triple-bottom line, the community the firms operate.

To respond to the call and to show their responsibility to the environment and the society, while earning profit, family firms are also put in a vague situation to respond. Because, family firms need to be more open to global calls and events, which can jeopardize their family name, tradition, and values. Therefore, they need smart enough leadership that will balance between the family values and the triple-bottom-line approach, which will lead them to achieve long-term sustainability in foreign countries in all these three areas (planet, people, and profit) (Fig. 8.1).

However, it is not an unbearable challenge for family firms because by their nature they are very much respectful of their surroundings even more than non-family firms. To successfully respond to this challenge, a family firm needs to accept the new market rules based on this trio-interplay between environment, society, and the firm.

Followed by introduction, in this chapter we introduce the nature of global operations, and then all three aspects of the triple-bottom-line approach as family firms go for global operations. Finally, the chapter provides the final view on the implications of global operations of family firms.

[4] United Nations: 17 Sustainable Development Goals: Available at: https://sdgs.un.org/goals. Accessed on December 12, 2023.

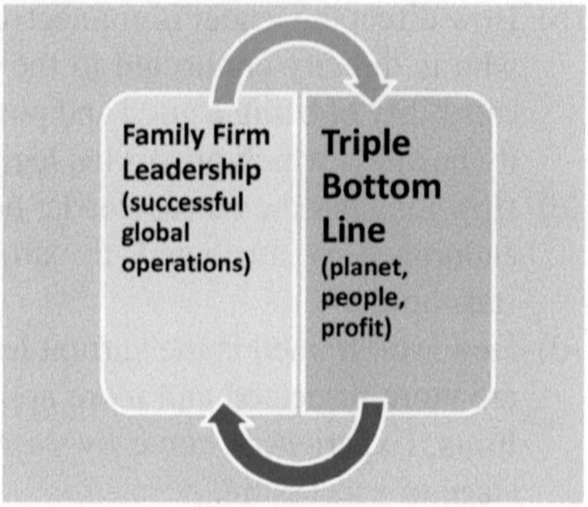

Fig. 8.1 Balanced approach of family firms' leadership to triple-bottom line. Source: Authors

8.2 Nature of Global Operations and Triple-Bottom Line

As discussed earlier in previous chapters, family firms differ from non-family ones in many ways, regardless of their size (Chang et al., 2022; Debicki et al., 2016; Lumpkin & Bacq, 2022; Ramadani et al., 2020; Smajić et al., 2022; Zellweger & Astrachan, 2008). Being different on the basis of family history that preceded the business, family firms' owners (managers or CEOs), carefully find the way to operate locally as well as globally. They cautiously choose partners in business (Pant & Rajadhyaksha, 1996) to preserve the credibility and reputation, built on a social image through years and generations, in the community and abroad.

For instance, Pant and Rajadhyaksha (1996) introduce the way that works if a family firm looks for a partnership in an Asian cultural context. The potential alliances in the Asian context may expect the following leadership pillar-check points, which are subject to a type of alliance (pp. 813–817):

- Ensure that the current leader is stable
- How stable are succession processes?
- How ready is the next generation?

Let us see what each of the above implies if family firms would do business partnerships or alliances in Asian region.

8.2.1 Check Number 1: Stability of the Current Leader

A strong leader in a family firm should have strong and authoritative attributes so that potential alliances have trust because the partnership will depend on the leadership qualities. The following factors impact the leadership stability:

(a) How much of the business is owned by the family leader? The weakness of the leader is regarded as if he or she does not possess the majority of ownership in the firm. If does, then it shows the authority and power in the family business.

(b) How direct the leader is connected to the family business's founder? The leader who is directly connected to the founder of the business bears the glory. The capability of being a good and powerful leader that has a true legacy in the family business which proves his legitimate authority.
(c) How long has the current leader been in the current position? Longer leadership endurance is attributed to stronger legitimacy for internal and external stakeholders.
(d) How well trained is the current leader? Well-trained leaders in the family firms are more respected and more appealed to make partnerships with other family firms. Experiences come by years and it is not possible to get it in the early stage of succession.

8.2.2 Check Number 2: Stability of the Succession Process

According to Pant and Rajadhyaksha (1996), there are several factors putting stability of the succession path unclear and unstable to potential cooperation. These are:

(a) Number of potential successors in the family firm. Different levels of managers and family members in the family firm think of potential succession opportunities, which for partners seems unstable leadership which can ruin the future business activities.
(b) Hierarchy of potential leaders. Ambiguous positions in family firms give a confusing impression and a bad image of partnership and cooperation. More clear hierarchy gives a positive hope for the internal and external future cooperation, which makes family firms prospective.
(c) Credibility of the selected leader. The credibility of the selected leader is contingent on different factors like how much he or she is linked to the founder, the shares, experience, training, and other similar factors.

8.2.3 Check Number 3: Readiness of the Next Generation (Successors)

It has been reported that succession in family firms is a big challenge for the business survival beyond the third generation. In family business field, this topic has always been hot and important (De Massis et al., 2008) because the question is how long the family entity will go on. Here are some takeaways if a family firm plans to have some partnerships in Asia.

(a) Training. Later generations of leaders in family firms are in a much better position than earlier generations. Specialized training in family firms provides better credentials to the family successor. Thus, in Asian cultural context, this plays an important role. An entrepreneur in a family firm who has specialized in a certain field is a big plus to his or her leadership (Barringer & Ireland, 2019; Longenecker et al., 2017).

(b) Business experience. It implies that some family firms' leaders are kin to prepare the next generation to lead the businesses. Some of these leaders give to their successors early responsibilities within the firms while others don't see it inevitable. However, Early life experiences play a crucial role in shaping the beliefs, attitudes, and career aspirations of family business heirs (Liu et al., 2024; Zellweger et al., 2011).

The above arguments provide a good path for how to succeed in global operations in this part of the world. Of course, some other factors could be taken into consideration because the above suggestions seem to be only one side of the coin. Other pathways may require how much the family business is involved in the implementation of the triple-bottom line to achieve long-term sustainability.

8.3 Other Schemes for Family Businesses Going Global

We looked at how family firms locate important crossroads to support international business partnerships with other businesses of a similar nature in the last topic. It appears that the rational principles that underpin all businesses, the pursuit of stability and trust to establish enduring cooperative relationships, are the foundation of these checkpoints.

Metsola et al. (2020) proposed an alternative internationalization process model. They integrated two types of studies, variance-based studies (antecedents, moderators, and outcomes), and process-based ones that focus on the dynamics of pre-entry, entry, and post-entry stages. Additionally, this model uses international involvement (internationalization/export propensity and intensity) and the time.

The model provides a holistic view of FB internationalization, highlighting the dynamic capabilities and liabilities that influence the journey of FBs in the global market. The model outlines four different pathways of internationalization (Fig. 8.2):

- **Unrealized or sporadic**
- **Incremental**
- **Born-global**
- **Born-again global**

The pace, speed, and scope of internationalization are reflected in these pathways. Important components are the *capabilities* (benefits for internationalization)

Fig. 8.2 Internationalization pathways of family firms: Global context. Source: Metsola et al. (2020), pp. 5–7

and *liabilities* (hindrances to internationalization) unique to family businesses. According to the model, capacities become more important later on (high internationalization intensity), whereas liabilities are more significant early on (low internationalization inclination).

Capabilities encompass both operational (exploitation of firm-specific advantages) and dynamic (adaptation to changing environments) aspects. *Liabilities* often relate to family ownership and management issues, such as risk aversion and centralized decision-making. In contrast, capabilities include factors like new generation involvement, human capital development, and focus on R&D.

The model suggests that successful family business internationalization involves minimizing liabilities before internationalization and capitalizing on capabilities during the process. Different pathways demonstrate how family businesses can navigate these stages.

The model also explores the firm-level economic and non-economic goals of family firms in their internationalization efforts. It suggests that as family firms become more internationally involved, their focus shifts from non-economic to economic goals, in line with the increasing significance of capabilities over liabilities.

For example, in *incremental pathways*, pre-entry and entry stages are cautious due to liabilities, but post-entry stages might leverage capabilities. *Born-again-global* pathways indicate a rapid shift to capability-driven internationalization following critical events like management changes.

In summary, this analysis presents a nuanced understanding of FB internationalization, emphasizing the importance of *balancing liabilities and capabilities* and recognizing the dynamic nature of this process across different stages and under varying conditions. The model considers the dynamic nature of family firms, acknowledging that new generations and external managers can significantly influence internationalization practices and resource commitments.

The above models presented make a good basis for family firms to establish global operations and to embed the triple-bottom line. It could be a two-edged sword for family firms. However, as the global world is headed toward a safe, secure, and valuable environment, family firms take this opportunity to upgrade their strategic positions at the global level. The following section considers global operations through the 3Ps (planet, people, and profit) as the triple-bottom-line approach.

An additional model that proposes survival and sustainability in "black swan"(Rothaermel, 2017) events, like recent pandemic, was proposed by Palalic et al. (2022) in which multinational companies find the way to long-term sustainability. The model is based on the term "think globally, act locally" (Dubos, 1981), which has been used for decades and yet it works till now.

In this model, they proposed that global businesses should have three imperatives in their corporate global strategy:

- Establishing their own reserve fund.
- Engaging in the local mutual funds' establishment with local industry players.
- Use the local networks as the average for their business activities.

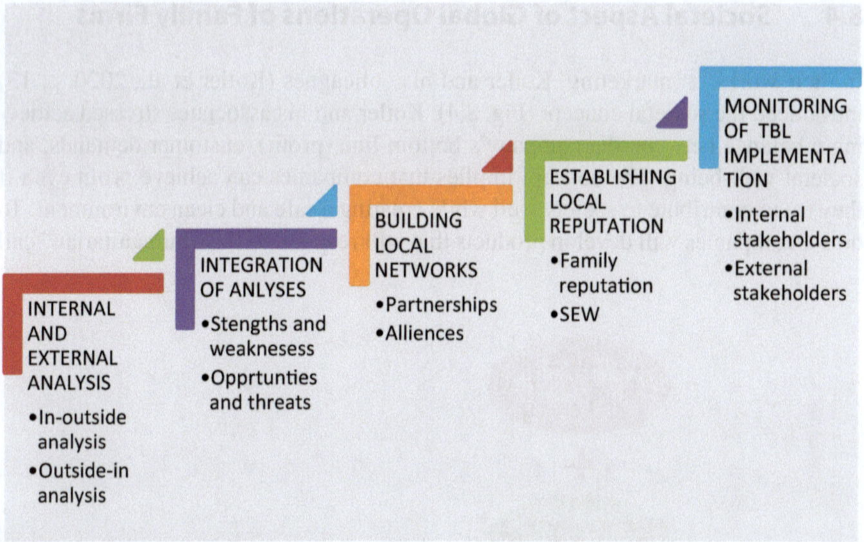

Fig. 8.3 Family firms going for global operations. Source: Authors based on Longenecker et al. (2017), Metsola et al. (2020), Palalic et al. (2022), Pant & Rajadhyaksha (1996)

To respond to the new world agenda, family firms need to "head up" and adopt the TBL approach in their corporate agenda for global operations. The following pathway can be of help to gain global attention and simultaneously implement the TBL approach, societal, economic, and profit agenda (Fig. 8.3):

1. Do internal and external analyses prior to entering a region or a country. Family firms should perform "outside-in" and "inside-out" analyses (Longenecker et al., 2017) in which they will realize the family firm's weaknesses and opportunities and whether they are capable or not to pursue the business in that region (pp. 75–79).
2. Conduct SWOT by integrating these two analyses. The analyses will show how ready is the family firm to undertake the global operations in the region it has targeted. If analyses show that the firm has competence and capabilities, it should think of forming a team for the operations.
3. Establish a global team in the host country and recruit people from the targeted region.
4. Work on establishing local networks (Palalic et al., 2022) or partnerships (de Farias et al., 2009) that will enhance a better business life. Inspect all possibilities and challenges of having a partnership or an alliance (Pant & Rajadhyaksha, 1996).
5. Create corporate values that will enhance the family firm's image in the targeted region as *friendly* and *positive.* The preservance of the socioemotional wealth (SEW) should be mandatory because of the historically positive image of the firm, which can help the new global subsidiary to help in establishing global credentials of the subsidiary.
6. Monitor all activities and make sure that all family firm's stakeholders (internal and external) implement the values embedded in the TBL approach.

8.4 Societal Aspect of Global Operations of Family Firms

In their works on marketing, Kotler and his colleagues (Kotler et al., 2020, p. 13) introduced the societal concept (Fig. 8.4). Kotler and his associates stressed achieving a balance between the company's bottom line (profit), customer demands, and societal well-being. The concept implies that companies can achieve profit even if they try to contribute to society well while creating a safe and clean environment. To do so, companies will develop products that will respond to this "humanitarian" call

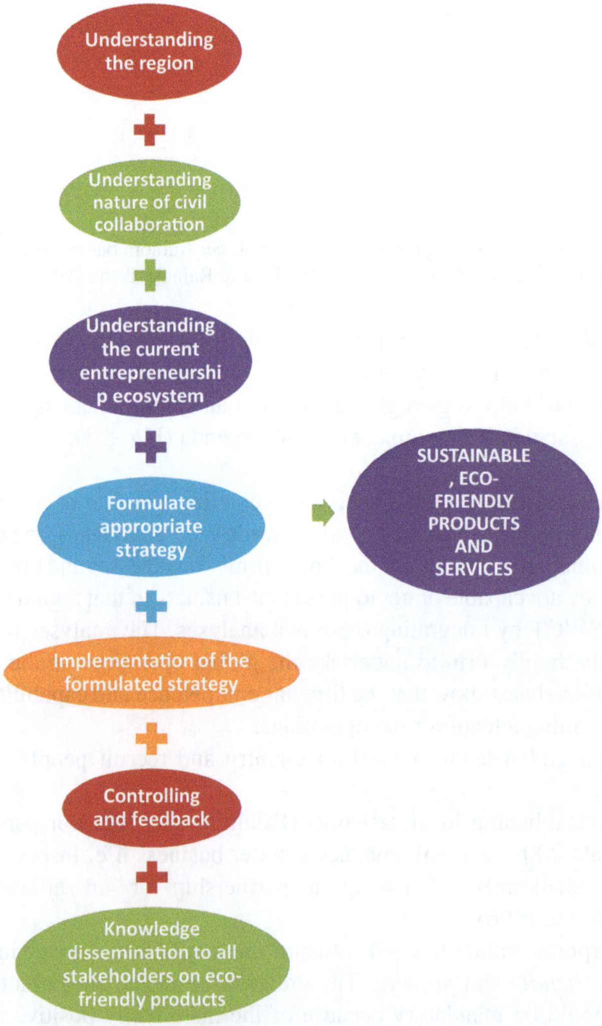

Fig. 8.4 Pathways toward sustainable, environmentally focused, products and services. Source: Authors, based on Sect. 8.5 (environmental aspect of global operations of family firms)

to save the planet, make society happy, and eventually feed customers with their desired needs, wants, and demands.

Similarly, family firms can use the triple-bottom line to enhance their global operations. Furthermore, the advantage in competition could be in danger if any of the game's stages are played improperly. Customers will purchase the products offered by the businesses, resulting in revenue for the business. Businesses, on the other hand, should address societal demands in a way that benefits the community, customers, and the business in an equitable manner. Although society has a big impact on consumers and can send a favorable or negative message about a company's performance, society does not directly affect a company's ability to make money. However, if the outcome is favorable (community plays in the firm's favor), new as well as loyal consumers are attracted.

Family firms will play a similar game in global operations. They will succeed within this aspect if they understand the business surroundings, like:

- If they understand the societal issues and problems in the region they operate.
- If they proactively search for an easier life in the society in the targeted region.
- If they produce products or provide services needed by the people and/or make life easier to the society in the targeted region.
- If they monitor and properly address the feedback on their business activities in the society.

The success of family firms in global operations is closely linked to their understanding and engagement with the societal context of the regions in which they operate.

In essence, family firms that are adapted to the social dynamics and needs of their business surroundings and that integrate this awareness into their business models, are better positioned to achieve sustainable success. This approach reflects a commitment not just to profitability but also to the principles of the TBL approach benefiting people and the planet, alongside their own economic interests.

8.5 Economic Aspect of Global Operations of Family Firms

Assuming that family firms in global settings create products or services that will resolve some problems from different perspectives of consumers, it is expected that the value in return will not be jeopardized. Yet, it will be a good milestone for long-term sustainability in economic terms. Moreover, family firms should look the global operations as an imperative (Patel et al., 2012) due to a variety of advantages that can assist in creating a new value to the family businesses.

In order to apply the "profit" component (Fuentes-Lombardo et al., 2023) of the triple-bottom line to their international (global) operations, family businesses should balance their other two aspects with a sustainable profit margin. It should be noted that to enhance the economic aspect and growth of the family firms (Fuentes-Lombardo et al., 2023) in the light of the TBL, family businesses should take advantage of their special qualities, like ability to build strong interpersonal ties in society and in-depth industry knowledge, to enhance productivity and effectiveness in international operations. Local partnerships can help in achieving a better economic outlook that will be sustained in commonly developed products and services that will serve society and customers (Gamble et al., 2021).

By doing this, they fulfill the comprehensive goal of the TBL and not only guarantee economic success (profit) but also make a beneficial contribution to the communities and environments in which they operate.

Ernst & Young (2021) reported that "family businesses in the 2021 Index collectively generated US$7.28 trillion in revenues, employed 24.1 million people and were distributed across 45 jurisdictions. Together they constitute the third largest economic contribution in the world (after the US and China) by revenue1—despite the global economy shrinking by 3.5% in 2020. They are vital to the future health and growth of every country's economic well-being during the post-pandemic recovery." This shows how much family businesses contribute to the overall economic outlook around the globe, besides achieving their primary objective of earning profit.

8.6 Environmental Aspect of Global Operations of Family Firms

One of the aims of family firms in global business settings is to improve overall global socio-economic status and well-being of people (Baù et al., 2021). From an environmental perspective, as discussed in Chap. 7, it is narrative that companies which pursue eco-friendly products are oriented toward a new agenda of sustainability, either they are attuned to environmentally sustainable goals (ESGs), sustainable development goals (SDGs), or triple-bottom-line approach. Positive impact on people and environment, actually they are contributing to overall well-being on this planet, as well as to the firm's long-term sustainability.

The foundation of today's industrial systems was and is still carbon-based energy systems, excessive non-sustainable resource exploitation, and a debt-based consumerist philosophy (Aguiñaga et al., 2018). It denotes that there is still a need to step in to study, observe, and concur with the region that lacks environmentally safe products and services.

Although some family firms act mindlessly regarding the SEW protection on the cost of poor business performance (Gómez-Mejía et al., 2007), many of them are very much mindful of having strong ties to the environment through their SEW (Dayan et al., 2019). Yet, their sustainable competitive advantage is based on being sincere and fair players in the environment (Dayan et al., 2019).

Being a global player in the business in someone's else "backyard" (in a foreign country) is very challenging. Family firms are resilient enough because they are mindful of finding way to adapt and adopt to changes in a dynamic environment.

To overcome this challenge of being in a foreign country, and to enhance their sustainable competitive advantage via TBL approach, to contribute to globe by providing safe and eco-friendly products and services, family businesses need to apply the following steps, Fig. 8.4:

Step 1: Understanding the region they operate. It means that is necessary to be familiar with social, economic, political, and environmental constraints (Aguiñaga et al., 2018).

Step 2: Understanding on how civil or public collaboration works. It is necessary to observe how to get connected with public organizations as potential partners in dissemination of the main objectives of the family firms (Aguiñaga et al., 2018).

Step 3: Understanding the entrepreneurship ecosystem of the region they want to enter. Some entrepreneurship are well-developed while others need other environmental actors to enter to enhance the current entrepreneurship ecosystem (Bichler et al., 2021).

Step 4: Formulate the appropriate strategy to respond previous three steps. Once family firms understand the overall situation and have gathered all necessary data, they can formulate a strategy to design business operations that will contribute to the environment through the sustainable, eco-friendly products and services. If the strategy is successful, the SEW will be enhanced, public image will be denoted among the people, and loyal customers will be attained.

Step 5: Implementing formulated strategy. Implementing the formulated strategy efficiently (using efficiently available resources) and effectively (setting the right or appropriate tasks to implement strategic goals).

Step 6. Controlling and feedback. Monitoring all activities regarding the strategy implementation and feedback.

Step 7: Disseminate knowledge on eco-friendly and transparent products and services in a transparent way (Bhaduri & Ha-Brookshire, 2011). Being honest, and transparent toward all stakeholders, especially consumers, strengthens the position of family firms in the region they operate. Consumers are kin to know about eco-friendly products and they are many times the ones who are environmentally aware (Reddy et al., 2023).

8.7 Implications of Triple-Bottom Approach of Family Firms' Global Operations

Globalization can be defined as free movement of goods, services, people, and capital across borders. Globalization had taken place many centuries ago before some continents were discovered and were known to the world. One of the greatest examples is the "silk road" which had connected the China or Far East with the West, Middle East, and North Africa. In this exchange of goods and people, the main pattern of globalization was export-import. Such an exchange used to take a long time from one exchange to another. Simply, the world had lived slowly.

Modern globalization took place in the late nineteenth century, immediately after the industrial revolution. It has been enhanced by the newest technology and communication, which accelerated all globalization processes and took different patterns. While the world has progressed in its use of natural and other resources, which is reflected in climate change, the new world agenda has called for slowing down the unsustainable use of natural and other resources. All individuals, institutions, and organizations are called to contribute to this call. So do the family businesses.

In this chapter, we introduced the global operations of family firms that respond to the above call, and implement the triple-bottom-line approach as their sustainable competitive advantage by balancing between profit (economic aspect), the planet (environmental aspect), and people (societal aspect).

Global triple-bottom-line agenda. Although the world is not free from unsustainable exploitation of natural resources, the global agenda calls for implementing environmental, societal, and economic perspectives for every business, big or small, family or non-family business. Responses to this call vary from one region to another. In the West, sustainable development goals or environmentally sustainable goals are, while some Eastern countries are using carbon-free environments and circular economies. Others are trying to implement some part of the global agenda while some trying to encompass all three aspects, planet, people, and profit. Being resilient and mindful in most cases, family businesses, especially the ones that have existed for decades if not centuries, incorporate the TBL agenda in their corporate and global business strategies.

Nature of family businesses internationalization. Internationalization of family businesses is subject to different factors that prevent or leverage the internationalization of their businesses. Liabilities, based on socioemotional wealth, in many cases prevent family firms to operate internationally or globally, due to non-economic (non-financial) goals. However, capabilities on other hand, leverage family operations at the international or global scale in a successful manner due to economic goals. Different pathways of internationalization of family firms are observed, like unrealized (sporadic), incremental, born-global, and being-again global.

Societal aspect of family firms in global operations. It is based on the previously introduced concept whereas family firms should take into consideration, besides traditional view of attaining profit, the needs of the society, and well-being of people. In the global arena, family firms need to be thoughtful enough to observe and adapt their business operations according to the societal trends, which are aimed to contribute to the socio-economic development of the region.

Economic aspect of family firms in global operations. Businesses exist for profit, and so do family businesses. When family firms are involved in implementation of the TBL approach, and provide appropriate eco-friendly products and services to the people, their brand image as family business increases. Consequently, being in touch with societal problems, and solving the problems by providing appropriate products, family firms will earn monetary value in the long term. A base of loyal customers will be established denoting the long-term sustainability.

Environmental aspect of family firms in global operations. Within the TBL approach, family firms secure their long-term sustainability by creating environmentally safe or eco-friendly products. Additionally, the firms respond to

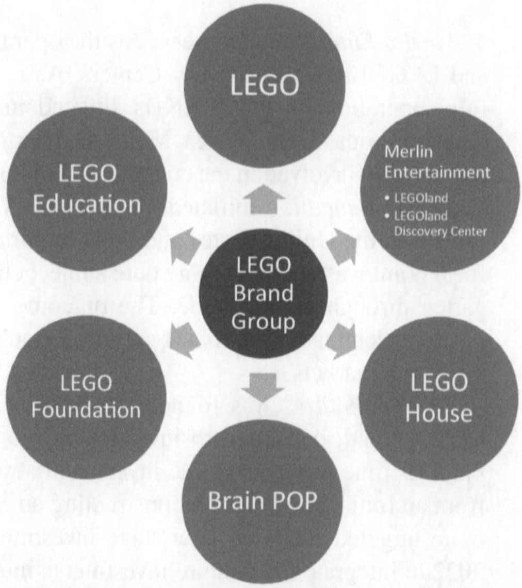

Fig. 8.5 LEGO Brand Group. Source: Figure developed based on official company report (KIRKBI, 2022)

environmental demands to save the environment in which we live. To provide sustainable eco-friendly products, the family firms go through seven different steps, as proposed in Fig. 8.4.

KIRKBI's Sustainable Global Operations[5]

About KIRKBI and LEGO Group

KIRKBI, a family-owned enterprise, together with LEGO Foundation is the main owner of LEGO Group. While LEGO Foundation holds 25%, the KIRKBI holds a commanding 75% stake in the LEGO Group. KIRKBI itself has a workforce of 200 employees, while the LEGO Group employs over 27,000 colleagues globally, having products available in over 100 countries. The LEGO Group organizes its operations through a network of 904 LEGO-branded stores. This includes 185 stores directly owned and managed by the LEGO Group, while the remaining 719 LEGO Certified Retail stores are operated by strategic partners. Having its production facilities strategically located in Denmark, the Czech Republic, Hungary, China, and Mexico, the company extends its presence beyond its Billund headquarters to key offices in the United States, the United Kingdom, China, and Singapore (Fig. 8.5).

[5] Ensar Mekic, Sultan Qaboos University. Based on the bibliography used.

Merlin Entertainments serves as the operator for both LEGOLAND Parks and LEGOLAND Discovery Centers. As of 2022, the company boasts ten fully operational LEGOLANDs situated in Billund, California, Germany, Dubai, Florida, Japan, Korea, Malaysia, New York, and Windsor. Additionally, it is actively involved in the construction of three new LEGOLANDs in China.

LEGO Education initiated a new strategy with an emphasis on fostering growth in the United States and most importantly international markets. The focal point was encompassing core subject classrooms spanning from kindergarten through eighth grade. The outcomes of this approach were notably positive, demonstrating robust growth not only in the United States but also in European markets.

LEGO Ventures was founded in 2018 as the venture capital arm of the LEGO brand, dedicated to investing in the future of education and digital play. As time progressed, the investment strategy of LEGO Ventures moved from an initial emphasis on constructing an early-stage venture portfolio to a more targeted focus on later-stage investments. A decision was reached in 2022 to integrate the venture investments into the broader investment portfolio of KIRKBI, leading to the gradual phasing out of LEGO Ventures as an independent entity.

BrainPOP was established in 1999 by former pediatrician and immunologist Avraham Kadar, MD. Its mission was to empower children in shaping both the world around them and within themselves. As a prominent digital education brand, BrainPOP's primary offerings consist of concise, animated movies designed to ignite children's innate curiosity while fostering the development of their knowledge and life skills. In October 2022, BrainPOP was acquired by KIRKBI which was KIRKBI's first complete acquisition of this magnitude. It served as a platform investment in the realm of digital learning to complement their existing portfolio of LEGO-branded entities.

LEGO Group has its stores in 13 countries in Asia Pacific, 22 countries in Europe, 6 countries in Middle East and Africa, 3 countries in North America, and 6 countries in South America. Even though the following figure does not show all Lego operations, it provides a basic insight about its worldwide presence.

Sustainability Orientation of LEGO Group

There is no doubt in LEGO Group's clear ambition to stand among the world's leading companies in terms of sustainability, embodying a triple-bottom-line approach. Back in 2015, the LEGO Group demonstrated its commitment by investing a substantial $150 million to establish the Sustainable Materials Center at its headquarters in Denmark. This pioneering center is dedicated to the exploration of innovative and sustainable materials for both packaging and the iconic LEGO bricks themselves. In the pursuit of more environmentally friendly alternatives, LEGO collaborates extensively with companies and institutions developing materials with a reduced environmental impact (LEGO, n.d.). It is worth noting that this commitment to sustainability was demonstrated before KIRKBI's signing of the Global Compact in 2019.

KIRKBI family keeps its promises

The foundation of KIRKBI's sustainability is deeply embedded in its Fundamentals and Promises which is formalized through the Code of Conduct. This way, the company establishes the minimum standards for upholding good corporate citizenship, spanning human rights, labor practices, environmental considerations, and anti-corruption measures. Crucially, the Code of Conduct extends its reach beyond KIRKBI's internal operations, encompassing broader business ecosystem in which it operates, such are the entire supply chain and other business partners. The company keeps the following promises:

- Planet promise
- Partner promise
- People and play promise.

One may notice that the company bases its promises on Triple-Bottom-Line (TBL) concept whereby it should achieve sustainability through responsible treatment of people, planet, and profits. Interestingly, KIRKBI, while not explicitly listing "profit" as one of its promises, envisions its enhancement by adhering to the following four commitments: the planet promise, the partner promise, the people promise, and the play promise.

KIRKBI adopts a responsible approach to the implementation of these promises, diligently monitoring each commitment through a set of key indicators. The progress and impact of these endeavors are transparently communicated through the publication of annual reports, which are available to the public on the company's official website. This commitment to transparency not only underscores KIRKBI's accountability but also serves as a testament to its dedication to the principles of sustainability.

Planet promise

KIRKBI has a comprehensive sustainable Real Estate strategy with identified key sustainability priorities crucial for managing its portfolios. These priorities notably encompass CO_2 emissions and natural resource management. Significant success in sustainability has been achieved at the KIRKBI headquarters in Billund, a building distinguished with a DGNB Gold certification. In 2022 alone, a commendable 20% reduction in total energy consumption was realized, with on-site solar panels contributing nearly half of the electricity consumed. Moreover, the company has intensified its commitment to environment by significantly enhancing waste management practices. From a baseline recycling rate of 50% in 2020, there has been a remarkable improvement, with recycling rates surging to an impressive 79% in 2022, underscoring KIRKBI's dedication to minimizing its environmental footprint. Beyond energy and waste management, KIRKBI has undertaken a substantial initiative to augment biodiversity around its headquarters. Biologists are actively engaged in assessing the changes in plant and insect levels, demonstrating a holistic approach to sustainability that extends beyond traditional metrics and incorporates the broader ecological impact of its activities. Considering this, it is not surprising that LEGO Group invested a substantial $150 million to establish the Sustainable Materials Center at its headquarters in Denmark.

Partner Promise

KIRKBI is committed to fostering mutual long-term value through proactive ownership that aligns with elevated Environmental, Social, and Governance (ESG) standards. Good evidence of this commitment is Armacell, one of KIRKBI's Long-term Equity investments. Here, KIRKBI assumes a leadership role as the chairperson of the company's Sustainability Committee, a position established by the Supervisory Board. The Sustainability Committee, functioning as the authoritative body for Armacell's sustainability initiatives, plays a pivotal role. It not only guides the internal Sustainability Steering Committee but also champions the strategic direction of Armacell's sustainability efforts. The committee prioritizes endeavors related to social and environmental aspects, ensuring a comprehensive and impactful approach to sustainability under KIRKBI's active ownership.

People and Play Promise

KIRKBI introduced a unique opportunity for KIRKBI employees, granting them the chance to allocate up to two working days annually to engage in the Local Community Engagement program. This program, a collaborative effort between the LEGO Group, the LEGO Foundation, and local charities, is designed to facilitate employee participation in events aimed at inspiring and fostering the development of children through play. By actively involving its employees in these initiatives, KIRKBI not only contributes to the well-being and growth of children in local communities but also gains valuable inspiration from its most important stakeholders—the children themselves. This program underscores the company's commitment to making a positive impact beyond our corporate boundaries and emphasizes the significance of play in nurturing the potential of future generations.

Discussion Questions:

1. How KIRKBI's dedication to sustainability affect the companies under its ownership?
2. Could you establish a clear connection between each of the four KIRKBI's promises and each individual element of the TBL approach?
3. Is global presence of LEGO Group challenge or push factor for sustainability?

How would you describe KIRKBI's global impact?

Bibliography used

KIRKBI. (2022). Retrieved December 2023, from www.kirkbi.com: https://www.kirkbi.com/about/annual-reports/

Google Maps. (2023, December). *Google Maps*. Retrieved from https://www.google.com/maps

LEGO. (n.d.). *www.lego.com*. Retrieved December 2023, from https://www.lego.com/en-us/service/help/facts_about_us/facts_about_us/sustainable-materials-research-kA009000001dbqpCAA

Questions for Discussion

- Define the globalization.
- Discuss international and global operations. Any differences? Provide justifications with your answers.
- How can family businesses achieve sustainable competitive advantage in global operations using TBL approach?
- Discuss possible challenges for family businesses in global operations and how they can avoid failure regarding TBL approach.

Additional Readings

- Lobo, C. A., Fernandes, C., Ferreira, J., Veiga, P. M., & Gerschewski, S. (2023). The Determinants of International Performance for Family Firms: Understanding the Effects of Resources, Capabilities, and Market Orientation. *Entrepreneurship Research Journal*, *13*(3), 773–811. https://doi.org/10.1515/erj-2021-0334
- Patel, V. K., Pieper, T. M., & Hair, J. F. (2012). The global family business: Challenges and drivers for cross-border growth. *Business Horizons*, *55*(3), 231–239. https://doi.org/10.1016/j.bushor.2012.01.002
- Sánchez-Chaparro, T., Soler-Vicén, M. Á., & Gómez-Frías, V. (2022). Be good and look good: Communicating the triple bottom line through corporate websites. *Journal of Business Research*, *144*, 136–145. https://doi.org/10.1016/j.jbusres.2022.01.089

Suggested Activities

- Select five family firms in your country that have their operations outside the country.
- Analyze the above family firms and discuss their internationalization.
- Based on the above firms, try to distinguish which one(s) is international and which one(s) is a global player.
- Based on the above information, what would you recommend to each of them in regard to global operations and TBL approach?

Keywords and Concepts

- Globalization
- Glocal
- Internationalization
- Unrealized or sporadic internationalization
- Incremental internationalization
- Born-global internationalization
- Born-again-global internationalization
- Outside-in analysis
- Inside-out analysis

References

Websites

Ernst & Young (EY). (2021). Available at: https://www.ey.com/en_om/family-enterprise/how-the-worlds-largest-family-businesses-are-proving-their-resilience Accessed on 19 December 2023.

http://millsproperty.com/about-3/, Accessed on 27.02.2020.

https://fambus.org/mills-property-management/ Accessed on 27.02.2020.

The Family Business Consulting Group. Available at: https://www.thefbcg.com/Secrecy%2D%2DA-Family-Business-Vice/. Accessed on 21.02.2020.

Aguiñaga, E., Henriques, I., Scheel, C., & Scheel, A. (2018). Building resilience: A self-sustainable community approach to the triple bottom line. *Journal of Cleaner Production, 173*, 186–196. https://doi.org/10.1016/j.jclepro.2017.01.094

Barringer, B. R., & Ireland, R. D. (2019). *Entrepreneurship: Successfully launching new ventures* (6, Global edition ed.). Pearson.

Baù, M., Block, J., Discua Cruz, A., & Naldi, L. (2021). Bridging locality and internationalization – A research agenda on the sustainable development of family firms. *Entrepreneurship and Regional Development, 33*(7–8), 477–492. https://doi.org/10.1080/08985626.2021.1925846

Bhaduri, G., & Ha-Brookshire, J. E. (2011). Do transparent business practices pay? Exploration of transparency and consumer purchase intention. *Clothing and Textiles Research Journal, 29*(2), 135–149. https://doi.org/10.1177/0887302x11407910

Bichler, B. F., Kallmuenzer, A., Peters, M., Petry, T., & Clauss, T. (2021). Regional entrepreneurial ecosystems: How family firm embeddedness triggers ecosystem development. *Review of Managerial Science, 16*(1), 15–44. https://doi.org/10.1007/s11846-020-00434-9

Chang, E. P. C., Zare, S., & Ramadani, V. (2022). How a larger family business is different from a non-family one? *Journal of Business Research, 139*, 292–302. https://doi.org/10.1016/j.jbusres.2021.09.060

Dayan, M., Ng, P. Y., & Ndubisi, N. O. (2019). Mindfulness, socioemotional wealth, and environmental strategy of family businesses. *Business Strategy and the Environment, 28*(3), 466–481. https://doi.org/10.1002/bse.2222

de Farias, S. A., Nataraajan, R., & Kovacs, E. P. (2009). Global business partnering among family-owned enterprises. *Journal of Business Research, 62*(6), 667–672. https://doi.org/10.1016/j.jbusres.2008.02.003

De Massis, A., Chua, J. H., & Chrisman, J. J. (2008). Factors preventing intra-family succession. *Family Business Review, 21*(2), 183–199. https://doi.org/10.1111/j.1741-6248.2008.00118.x

Debicki, B. J., Kellermanns, F. W., Chrisman, J. J., Pearson, A. W., & Spencer, B. A. (2016). Development of a socioemotional wealth importance (SEWi) scale for family firm research. *Journal of Family Business Strategy, 7*(1), 47–57. https://doi.org/10.1016/j.jfbs.2016.01.002

Fuentes-Lombardo, G., Sanchez-Famoso, V., & Cano-Rubio, M. (2023). Reasons for internationalisation of family business. *Journal of Family Business Strategy, 14*(4). https://doi.org/10.1016/j.jfbs.2023.100571

Gamble, J. R., Clinton, E., & Díaz-Moriana, V. (2021). Broadening the business model construct: Exploring how family-owned SMEs co-create value with external stakeholders. *Journal of Business Research, 130*, 646–657. https://doi.org/10.1016/j.jbusres.2020.03.034

Gómez-Mejía, L. R., Haynes, K. T., Núñez-Nickel, M., Jacobson, K. J. L., & Moyano-Fuentes, J. (2007). Socioemotional wealth and business risks in family-controlled firms: Evidence from Spanish olive oil mills. *Administrative Science Quarterly, 52*(1), 106–137. https://doi.org/10.2189/asqu.52.1.106

Kotler, P., Armstrong., Harris, L. C., & HE, H. (2020). Principles of Marketing (17TH ED). UK: Pearson.

Liu, P. C. Y., Zhu, F., & Wang, J. (2024). The apple doesn't fall far from the tree: Parenting styles and its effects on family business succession intentions. *Journal of Business Research, 172*. https://doi.org/10.1016/j.jbusres.2023.114429

Longenecker, J. G., Petty, J. W., Palich, L. E., & Hoy, F. (2017). *Small business management* (18th ed.). Cengage Learning.

Lumpkin, G. T., & Bacq, S. (2022). Family business, community embeddedness, and civic wealth creation. *Journal of Family Business Strategy, 13*(2). https://doi.org/10.1016/j.jfbs.2021.100469

Metsola, J., Leppäaho, T., Paavilainen-Mäntymäki, E., & Plakoyiannaki, E. (2020). Process in family business internationalisation: The state of the art and ways forward. *International Business Review, 29*(2). https://doi.org/10.1016/j.ibusrev.2020.101665

Palalic, R., Hisrich, R. D., Dana, L. P., & Ramadani, V. (2022). Sustainability of global and international business operations during the adversity and hardship. *Heritage and Sustainable Development, 4*(1), 18–26. https://doi.org/10.37868/hsd.v4i1.78

Pant, P. N., & Rajadhyaksha, V. G. (1996). Partnership with an Asian family: What every multinational company should know. *Long Range Planning, 29*(6), 812–820.

Patel, V. K., Pieper, T. M., & Hair, J. F. (2012). The global family business: Challenges and drivers for cross-border growth. *Business Horizons, 55*(3), 231–239. https://doi.org/10.1016/j.bushor.2012.01.002

Ramadani, V., Memili, E., Chang, E. P., & Palalić, R. (2020). *Entrepreneurial family businesses: Innovation, governance and succession* (1st ed.). Springer.

Reddy, K. P., Chandu, V., Srilakshmi, S., Thagaram, E., Sahyaja, C., & Osei, B. (2023). Consumers perception on green marketing towards eco-friendly fast moving consumer goods. *International Journal of Engineering Business Management, 15*. https://doi.org/10.1177/18479790231170962

Rothaermel, F. (2017). *Strategic management* (3rd ed.). Mc Graw Hill Education.

Smajić, H., Palalić, R., & Ahmad, N. (2022). Future perspective of socioemotional wealth (SEW) in family businesses. *Journal of Family Business Management*. https://doi.org/10.1108/jfbm-05-2022-0070

Zellweger, T., Sieger, P., & Halter, F. (2011). Should I stay or should I go? Career choice intentions of students with family business background. *Journal of Business Venturing, 26*(5), 521–536. https://doi.org/10.1016/j.jbusvent.2010.04.001

Zellweger, T. M., & Astrachan, J. H. (2008). Building a family firm image: how family firms capitalize on their family ties. *Family Business Review, XXI*(3), 347–363.

9 Toward the Long-Term Sustainability Through Leadership in Family Businesses

Learning Objectives
After reading this chapter, you should be able to:

1. Define leadership aspirations in family businesses.
2. Define TBL leadership in family businesses.
3. Address the importance of family business leadership in TBL approach.
4. Understand the long-term role of TBL leadership in family firms.
5. Understand the family firm leaders' role in future agenda for future generations in family firms.

Profile: Al Khonji Holding Group
The ongoing Al Khonji Spark of 102 years was lightened up by the brothers, Mohammed and Ahmed Al Khonji, in 1920. They have paved their way to be one of Oman's deep-rooted reputable family conglomerates today, from a small maritime trading company located in ancient Muttrah Souk. The specialization of the company in that period was in shipping and dhow materials. Following that period, the company was led by Abdullah Mohammed Al Khonji—the second generation. During their management tenure, they have expanded the company by getting into a larger scale of trading, wherein they started importing a new set of products, which included radios, batteries, and boats as well as the export of foodstuff including dry dates. In addition, Al Khonji company was the first company in Oman to be granted the permission to establish an electricity company, which was constructed in Muttrah. Following the same pattern, the family business experienced ongoing

R. Palalić et al., *Sustainable Family Business*, Springer Texts in Business and Economics, https://doi.org/10.1007/978-3-031-57999-8_9

developments and expansions, including the construction of Oman's biggest showroom in 1960, in Muttrah.

The showroom has many products displayed for sale, ranging from building materials to spare parts, electricity and water pumps, and lubricants. However, the family business was not able to possess its official registration until 1971, when the Company Registration concept was introduced by his late majesty Sultan Qaboos. After forming their company officially, they have expanded to the business by which they are most recognized, "real estate." In addition, in order to diversify their business, they have also expanded into other sectors, which are services, trading, projects, and investment sector.

The family business was initially managed by the family members until 2017, and when there was a change of management and since then Michael E Hansen is the leader of the company. Upon which, Mr. Michael has worked together with the team and looked at what they had as a whole portfolio and suggested that "we need to have a strategic plan and to focus on the core business, by allocating our existing resources properly, so that we can deliver certain objectives related to growth and productivity, hence be more efficient." Besides, they were focused to invest the money earned from the business in expanding the holding group. Moreover, a proportion of the money was invested in MSX, UAE and outside the Gulf region in order to implement risk management and achieve price growth and diversification.

The three pillars that assisted Al Khonji group to sustain its competitive advantage and thrive through difficulties are future-oriented vision, great team and strong balance sheet in terms of cash. Through looking to the future, they were able to see when and how the opportunity would arise and what should be changed to exploit it. The great team, in which every individual has his own strengths and unique ability to come up with ideas to adopt and keep outperforming, besides having a strong personal responsibility toward their professional obligations. A strong balance sheet in terms of cash has assisted the company to further grow.

During COVID-19, most of the companies were suffering all over the world, except Al Khonji group. Al Khonji group viewed the situation as a growth opportunity, and the cash earned from the tenants and investments, enabled them to acquire the intended potential companies, which eventually gave them growth.

The company believes that attaining a sustainable competitive advantage requires focusing not only on achieving a significant financial position but also focusing and meeting their social and environmental responsibilities. This is clearly supported by the company's vision, which is "To be admired as a leading family business group, helping to build a better future by making a significant economic contribution and social impact to the communities and in the activities in which we participate".

The vision of the company addresses the concept of Triple-Bottom Line (TBL). The TBL's argument is that the trend of conducting business has evolved from pure profit and looking at the financial success only, to caring for society and the environment as a whole. With the leadership of the manager and the participation of its different departments including the Human Resources Management Department (HRM), the company attempts to maximize the efforts in developing an outperforming company by instilling the TBL concepts within its operations and policies. Furthermore, their deep recognition of Oman being a part of the world as a whole, and them being a part of Oman, led them to the fact that any environmental distortion made due to the company operations, eventually will affect them. In addition, any challenges witnessed by the world, will affect Oman, and definitely would affect them. The company's efforts in meeting its obligation and maintaining **environmental sustainability**, is clearly witnessed in its various conducted activities. For instance, their implementation of the *enterprise resource planning system* enables them to conduct the required business transactions within the departments electronically, in order to decrease paperwork and encourage individuals to go green by reducing the amount of paper used in their transactions. Moreover, the HRM has contracts with paper recycling firms to recycle company's utilized papers on a regular basis. In addition, they emphasize training the employees to follow the recommended practices that minimize the potential *electrical waste*. Hence, they always make sure that they do not simply throw the gadget or the PC in the trash but dispose of it carefully to avoid harming the environment. Moving to the second focus of Al Khonji group which is meeting their **social responsibilities**. The company, including the HRM emphasizes on the *training and development of the employees*; they ensure that their employees are appropriately trained so that they do not injure themselves or the environment when doing their tasks and activities that require the use of chemicals, repair equipment, or other tools. Furthermore, *Medical insurance* is another illustration of the HRM role in meeting the company's social responsibilities. The company provides excellent health care to its employees, by ensuring that they are all protected by a medical insurance to get instant treatment and return to the company healthy. Besides this, AL Khonji *plays a vital role in helping the people in need* by providing zakat and donating to them directly or through reputable charities, as they do in Ramadan and in periods of cyclones, including the Shaheen cyclone. In addition, they have also helped the society by building housing units for low-income people and building a dialysis unit in Al Amirat. Lastly moving to the core focus of any business which is generating profits and maximizing the company's value. Al Khonji's efforts in achieving its objectives related to **financial aspect** are clearly witnessed throughout Al Khonji's journey. HRM follows a precise procedure when recruiting the employees, by *requiring certain abilities, skills, and knowledge* to better serve the position offered, which in turn will contribute to higher profits earned relative to the cost incurred by the company. Furthermore, the *training provided by the HRM* also results in enhancing the employees' performance to accomplish their respective duties and responsibilities effectively and efficiently.

AL Khonji holding group is the best example of a successful family business that was started by a small business long ago and through their high determination and effective strategy, they achieved their goal of being one of Oman's leading diversified family holding companies that we know today. In addition, Al Khonji group's supportive vision and strong belief to consider the environmental and social aspects along with the financial aspect, has assisted the company to achieve a sustainable competitive advantage. In addition, the presented case about the Al Khonji group illustrates the importance of having a future-oriented vision, a great team and a strong balance sheet in terms of cash, in discovering business opportunities and exploiting them accordingly to achieve business growth and expansion.

Questions for discussion

1. How do you see this family enterprise in the light of TBL approach?
2. Do you think that the firm should be focused on one of the TBL pillars, either, environment (planet), society (people), or economy (profit)? Why? why not?
3. Based on the information from the case, what would you suggest to the family firm in regard to TBL approach?
4. Is it possible that a family firm like the one from the case, implement fully the TBL agenda?

Bibliography

Al Khonji Group. Al Khonji. (n.d.). Retrieved April 13, 2022, from https://www.alkhonjigroup.com/

Personal Interview: An Interview was conducted on March 3rd, 2022, with Al Khonji group's CEO (Michael E Hansen), Head of HR (Ibrahim Al Amri), IT Specialist (Reham Yassin), and Accounting Specialist (Afra)

9.1 Introduction

Family business' *emotional history* makes that firm a great difference from ordinary business in the market. Emotional moments that are derived from events during the young age of the family firm are transmitted from one generation to another. Many times these events are used to make strategic decisions to preserve non-financial values in the family owned (socioemotional wealth-SEW) (Ramadani et al., 2020a, b, c). Moreover, the emotional history can be used even to transmit a good tradition of an entrepreneurial family business venture to next generation, embracing a "transgenerational entrepreneurship" (Basco et al., 2019).

Therefore, the current leaders in family businesses have quite challenges in the way that they need to establish a proactive family enterprise with a clear future agenda to be profitable and sustainable while at the same time the leaders need to preserve the family business venture's tradition. Family business leaders "play between two fires," and should not be "burned" by any. In other words, they need to have a balanced approach. That is why "big fathers" of the family business carefully chose a successor among family firm members.

Having in mind the complexity of leadership in family businesses, leaders are challenged by external factors, the general environment, and internal factors—the emotional tradition of the firm. Moreover, the current world agenda is to set up a better world for all, sets a level up of challenge for the family firm's leaders.

Therefore, in this chapter we try to bring a better understanding of family firm's leaders surrounded by the universal (world) strategic development goals (SDGs; 17 goals) set by the world organization (United Nations-UN).[1] These goals are reflected in the "world agenda," the 5Ps: "people, planned, prosperity, peace, and partnership."[2]

As the firms are taking the part in this agenda, hence, the family firm's take part too. For their leaders, they need to revisit a new vision and mission that will help the people and planet to revitalize their natural need to establish a "symbiotic life" (Etemad et al., 2001) while earning profit on a newly set vision and vision.

After the introduction, this chapter brings the importance of the role of leadership of family ventures in a newly world-set agenda for sustainability, through people, planet, and profit. We complete the chapter with implications of TBL in family businesses.

9.2 Nature of Leadership in Family Businesses

Leadership is defined as impactful influence which makes people willingly committed to achieving their goals on a daily basis. Moreover, is an ongoing process in which all parties satisfactorily receive benefits in their mutual relationship as per their formal and informal roles in a private-owned entity.

[1] United Nations: Sustainable Development Goals. Available at https://sdgs.un.org/goals, accessed on September 29 2023.

[2] United Nations: World Agenda for 2030. Available at https://sdgs.un.org/2030agenda, accessed on September 29, 2023.

9.2.1 Uniqueness, Complexity, and Ambidexterity of Leadership in Family-Owned Firms

Compared to conventional businesses, leadership in a family business is more concerned about owner-leadership dynamics because the family business, as it grows, it becomes very complex (Longenecker et al., 2017; Ramadani et al., 2020a, b, c). The intense complexity is even underscored when it comes to successors. The propensity of the individual interest in the family business is strong along all development stages of the family business. Therefore, such competing interests among family members can create uncertainty in a family business leadership. To prevent such uncertain negative consequences, the leadership is paramount to appropriately channel all mentioned uncertainties.

Leaders in family businesses easily amortize negative effects of competing interests in the family businesses through a strong family corporate culture (Longenecker et al., 2017), which should be in line with the sustainable competitive advantage strategy (Rothaermel, 2017). Family-owned businesses have corporate culture imprinted from founders (owners) since their inception. Business founders are usually the source of innovative ways to respond to customer needs and build positive and long-term relationships. Acting as an exemplary, employees follow the role, the founder. That is how the competitive advantage is built and sustain because family business owners have a special (differentiated) way of keeping customers around and responding to their claims in a positive manner (Longenecker et al., 2017) as well as to business partners and clients.

The other side of the coin that looks shiny is the family members. Their commitment is one of the critical milestones for future succession of the family business. Pointing to the founder, the "big father," family members follow the exemplary role of the big father to preserve his commitment toward the business and his clients, customers, and business partners.

Such uniqueness of family businesses, and dynamics and complexity from another side, make family business ambidextrous that can play on different fields. For instance, family businesses simultaneously cope to sustain the firm competitive advantage via different strategies that respond to clients, customers, and partners successfully, and trials from the fact that family-owned firms are afraid of bad succession to next generations.

Nonetheless, the family firms keep this ambidextrous (Fig. 9.1) value for a few generations, because after third generation the family business becomes extremely complex in terms of ownership, decision-making, and growth. However, being an ambidextrous firm does not avoid being exposed to future challenges as explored in the next section.

9.2.2 Challenges of Ambidextrous Family Business Leadership

Family business is often described as agile and eager to respond to new challenges that will preserve the family tradition and values from one side, and increase

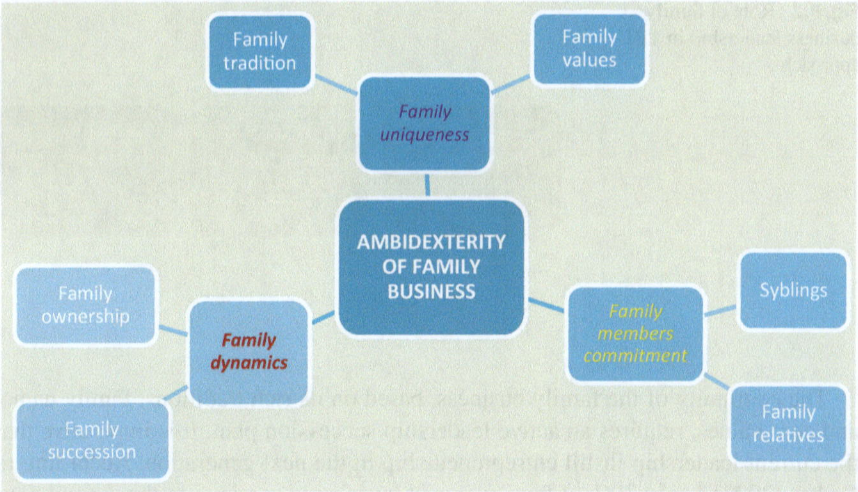

Fig. 9.1 Ambidexterity of the family businesses via family uniqueness, dynamics, and family members' commitment

Table 9.1 Challenges of ambidextrous family business leadership

Item	Description
Conflicts	Successful management of family members and cross-generational and siblings' conflicts
Family succession planning	Choosing the right person as a future successor of the family firm
Preserving tradition and values vs. professionalism	Managing overlap and balancing between professionalism and tradition and values in family firms
Strategy for family business innovation	Designing a strategy that will lead to innovation and sustainable performance

innovative products and services from another. The offset of these two sides of the scales (family values and tradition vs. innovation and growth) brings challenges that often could be turning points that may lead to different directions of the family business.

Common challenges that family-owned firms face in their growth and development are *conflicts* in family firms, *family succession planning*, *preserving tradition and values versus professionalism*, and *strategy for family business innovation* (Table 9.1).

Conflicts in family firms are a reality. However, family firm leaders should be smart enough to overcome and manage eventual disputes and disagreements (Ramadani et al., 2020a, b, c). Conflicts can spill over to new generations. Therefore, **succession planning** in the family firm is another challenge that leaders should count on and be ready to resolve as soon as possible so that the family firm leaders will successfully avoid further conflicts.

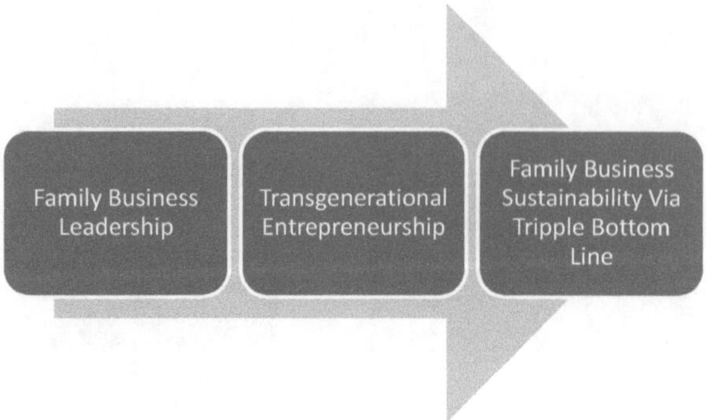

Fig. 9.2 Role of family business leadership in TBL approach

The continuity of the family business, based on its rich traditions, family name, and core values, requires an active leadership succession plan. It is imperative that the current leadership instill entrepreneurship in the next generation. According to Forbes (2023),[3] only 30% of family-owned businesses continue to the second generation, with only 12% reaching the third generation, and a mere 3% lasting through or beyond the fourth generation.

This strategic planning ensures that family wealth will translate into a lasting positive impact on the sector. In this way, family business leadership plays a key role in generating future generations with the skills and perspectives needed to assume leadership roles in the family business. Often referred to as succession planning, this process is a deliberate attempt to cultivate an entrepreneurial spirit in those who will succeed. By doing so, leaders ensure that the family business not only survives but thrives under the guidance of each successive generation.

This way of thinking recognizes that family business success is intricately tied to the intellect and ability of future charisma by preparing the next generation with an entrepreneurial mindset on family leadership paves the way for continuous improvement, innovation, and longevity.

Concretely, the emphasis on developing such an entrepreneurial mindset becomes a milestone in the strategic organization of the family business, giving it its legacy and relevance in the evolving entrepreneurial environment. This can be done through so-called "Successful Transgenerational Entrepreneurship Practices" (STEP) (Basco et al., 2019), which is predecessor of the transgenerational entrepreneurship concept (see Fig. 9.2). This concept, defined as the "processes through which a family uses and develops entrepreneurial mindsets and family influenced capabilities to create new streams of entrepreneurial, financial and social value across generations"

[3] Source: Forbes, 2023. Available at: https://www.forbes.com/sites/forbescoachescouncil/2020/03/18/strategies-for-family-business-success/?sh=3af6547e71c7. Accessed on August 20, 2023.

(Habbershon et al., 2010, p. 1). The concept is based on families and entrepreneurial orientation (Basco et al., 2019).

Thus, the family firm leaders need to select the right successor for leading the family business and put him or her in the succession planning development process.

A great example is of KIRKBI family-owned firm, with the fifth generation so far. One of the very significant activities of KIRKBI is an initiative of "The LEGO School." The primary objective of this program is to foster the development of their successor's distinct identities developing them at the same time both as individuals, but also as integral members of the family community. Simultaneously, it seeks to cultivate the skills necessary for them to assume their roles as future fifth-generation owners. This innovative approach is designed to unfold within a vibrant, secure, and playful environment. The LEGO School is an excellent example of a preparatory platform for the upcoming generation of owners in the family business.[4]

To have generational succession in the family firm, family firm founders' imperative is to transmit the entrepreneurial mindset to keep the business alive for next generations as entrepreneurial leadership. Some therefore argue that family longevity and intergenerational success in family firms are heavily influenced by two dimensions, familiness, and entrepreneurial orientation (Calabrò et al., 2022).

Another challenge that can destroy the family business's performance is **family tradition** and values that may somehow conflict with **professionalism**. This happens if the family firm takes many family members to enjoy family values so that nepotism or similar events are observed. Smart family business leaders should be always alerted of this possibility because the existence of the family business and its values and traditions fade over time. The leader in this case needs affirmative leadership that will integrate both the tradition and values while keeping the professionalism at a solid level.

If these two quite opposite challenges could be preserved and put under the control, then the family-owned business leader can with enthusiasm plan the innovation strategy that will keep the family firm competitive in the market. That is what the family firm needs to be sustainable, the strong leadership that knows where, how it will be headed toward long-term sustainability.

[4] Source: KIRKBI. (2022). Annual Report 2022. Retrieved December 2023, from www.kirkbi.com: https://www.kirkbi.com/about/annual-reports/

9.3 TBL Leadership in Family Businesses[5]

Certain family businesses choose to maintain a discreet stance when it comes to sharing their philanthropic endeavors with the wider community. This discretion may stem from cultural norms that discourage apparent displays of their family firm's humanitarian actions. Consequently, they may not publicly disclose their support for veterans or the families of fallen soldiers, or their commitment to assisting those facing joblessness or other urgent needs within their local community. For example, "Ugarak-Produkt, doo," a family-owned enterprise, is deeply involved in philanthropic initiatives within its local community. This family business puts a special emphasis on employing veterans and their family members, actively contributes to building homes for those in dire need, and extends job opportunities to individuals facing severe unemployment challenges. While such actions may not align with conventional marketing strategies, the guiding principle, as perceived by the "big father," is preserving dignity and credibility, refraining from extravagant displays that could potentially astray from the family's core values and the business's fundamental mission.

Source: Authors' interview with the owner of "Ugarak Produkt," LTD, November 2023.

If a family business has adopted the idea of triple-bottom-line approach, then the family business leadership should represent the perfect combination of economic well-being, social responsibility, and environmental stewardship. This philosophy recognizes that business success should be measured not only in economic terms (financially the profit) but also in terms of its impact on people and the planet. With a such mindset, family leaders play an important role in shaping simultaneously a sustainable and ethical corporate culture.

Given the triple commitment to profit (financial aspect), family business leaders seek profit along with long-term sustainability. Decisions are made considering not

[5] Source: A screenshot from the company's webpage: https://ugarak.ba/en/homepage-2-en/ Accessed on December 11, 2023.

only the immediate economic benefits but also the sustainability of the business and its stakeholders. Inherited an emotional aspect of the family firm through its family values and a focus on family legacy, family business leaders try to overcome the challenges, ensuring that the business remains financially viable for future generations.

The social aspect of the family firm leadership is that family business leaders put the well-being of their employees first, and foster a workplace culture that emphasizes diversity, equity, and inclusion. They understand the importance of building strong relationships with employees, customers, and the community.

This approach not only enhances the company's reputation but also contributes to social development. Family leaders often support charitable causes, support causes that align with their values and solve social challenges (Ramadani et al., 2020a, b, c). Additionally, family big fathers (founders of family business) go for an extra mile, solving the issues about war veterans and their families giving them priority while recruiting new employees (i.e., Ugarak Product LTD.) (Ramadani et al., 2020a, b, c).

The cornerstone of family business leadership is environmental responsibility. Surely, family firm leaders recognize the impact of business activities on the environment and actively implement sustainable practices. From energy-efficient operations to responsible supply chains, they strive to minimize the environmental impact of their operations.

Triple-bottom-line leadership needs cross-generational sustainability and longevity. However, the life of a family business in most cases goes up to three generations, unlike the Merck Group is being alive for 354 years (Basco et al., 2019), with 13 generations and employed 64,000 employees across 66 countries (Merck Group, 2023).

Naturally, family-owned leaders try to make family businesses alive throughout generations, which brings pride to the family name. To bring the business to next generations in which the triple-bottom line will be inherited and cherished throughout regenerations, the *big father* needs to be aware of necessity of the "transgenerational entrepreneurship," which will enhance the sustainability across generations. The "entrepreneurial mindset" seems to be very important (Ramadani et al., 2020a, b, c).

9.3.1 Transgenerational Entrepreneurship: A Way to Success and Longevity

Transgenerational entrepreneurship (Fig. 9.2) is explained as "a process through which a family uses and develops entrepreneurial mindsets and family influenced capabilities to create new streams of entrepreneurial, financial and social value across generations" (Habbershon et al., 2010, p. 1). The implication of transgenerational entrepreneurship suggests that the sustainability of a family firm leans on three key pillars: sustainability, entrepreneurial orientation (EO), its familiness (FAM), and the cultural context (CC) (Basco et al., 2019).

Figure 9.2 implies the cultivation of transgenerational entrepreneurship in family business (by family business leaders) to gain the long-term sustainability.

Furthermore, it is in the same line with TBL approach as it has it has components of the environment (planet), people (social responsibility), and profit (financial aspect) aspect. Moreover, transgenerational entrepreneurship nurtures the idea of the long-term sustainability of the family firm.

9.4 Implications and Future Agenda for TBL Leadership in Family Businesses

Family firms play a crucial role in the socio-economic development of nations. Despite the increasing focus on various aspects of family business in research, it is not studied as extensively as regular small and medium enterprises (SMEs). Some may view family businesses as less significant compared to SMEs, but it is essential to recognize that they are at least as important and contribute significantly to national development in various ways.

Leadership in family businesses requires attention to several aspects, and one key consideration is ambidexterity. This aspect is imperative for family-owned firms due to their unique nature, different from regular SMEs (Gómez-Mejía, et al., 2007, 2012). Family businesses often prioritize traditional values and the preservation of non-financial wealth, known as socioemotional wealth (SEW) (Debicki et al., 2016; Kalm & Gomez-Mejia, 2016). This commitment to preserving SEW can influence decision-making, sometimes leading to decisions that may not be the most financially optimal (Ahmad et al., 2022; Palalić & Smajić, 2022).

In essence, understanding the importance of family businesses alongside SMEs is crucial, and recognizing the need for ambidexterity in leadership helps navigate the unique challenges posed by the preservation of socioemotional wealth. Furthermore, while family firms must generate profits, it is essential to preserve the socioemotional wealth (SEW) component. This aspect sets them apart from SMEs, as highlighted by Berrone et al. (2012) and Zellweger and Astrachan (2008). Therefore, family businesses must embody ambidexterity in their operations.

9.4.1 Future Agenda for Family Business Leadership: TBL Perspective

Family businesses thrive on strong values and are guided by strong leadership. Leaders in family firms aim to ensure the longevity of the family business while at the same time maintaining the family name, history, and traditions. This commitment is crucial to maintaining a positive and respectable image in society. In family firms, leaders (owners), play a key role in influencing positive change and overall well-being in the community. Their responsibilities are various and thus complicate their role.

These leaders serve as influential individuals in society, contributing to positive change over the years. Their focus on inheritance, positive transitions, and successful succession sets family firms apart from SMEs. Specifically, leaders in family firms play a unique and lasting role in the business and its impact on society.

The recent global initiative (in 2020) led by the United Nations (UN) to save the planet, outlined in its Sustainable Development Goals (SDGs) (UN, 2023), is reshaping the market dynamics. With the UN setting conscientious goals that all firms and individuals should follow to contribute to saving the world, family firms have a unique opportunity to elevate their societal standing. They can do this by embracing the Triple-Bottom-Line (TBL) approach, aligning their goals with economic, social, and environmental factors.

To successfully implement the TBL approach, family firms need to craft a sustainable future agenda. This agenda, depicted in the figure, should include the following key elements.

1. Ambidexterity. Family firms should address the following points to achieve the ambidexterity:
   (a) Conflicts
   (b) Family succession planning
   (c) Preserving tradition and values vs. professionalism
   (d) Strategy for family business innovation
2. Successful Transgenerational Entrepreneurship Practices (STEP). This includes the following important aspects:
   (a) Transgenerational Entrepreneurship, which consists of:
      (i) Familiness
      (ii) Entrepreneurial orientation (EO)
      (iii) Cultural context (CC)

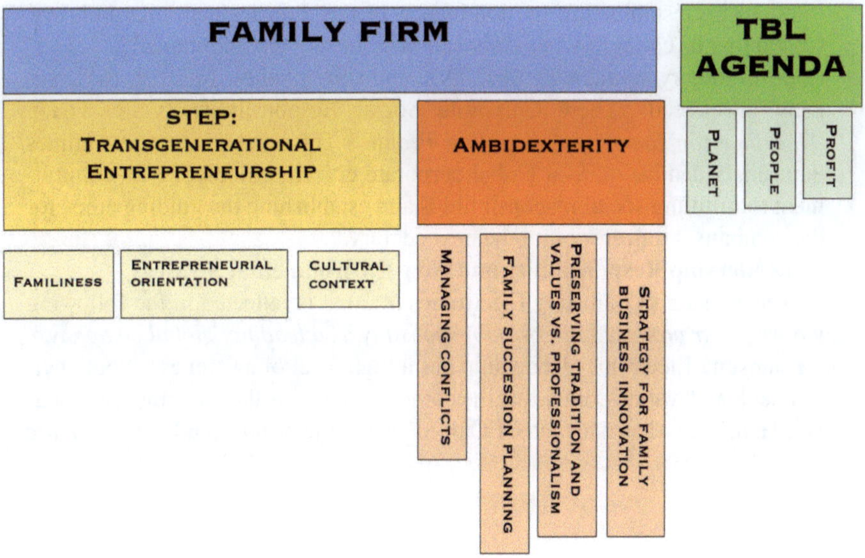

Fig. 9.3 Family firm leadership and TBL framework: A way to success

Case Study[6]

Green Leadership at Samsung Electronics

In 2004, Samsung held the notable rank of 3rd among the World's 250 Largest Family Businesses according to Family Business Magazine. During that period, the Lee family maintained control over approximately 22% of the ownership. The flagship division, Samsung Electronics, emerged as one of the world's leading manufacturers of computer memory chips and engaged in the production of various consumer electronics such as home electronics equipment, mobile phones, and microwave ovens. Beyond consumer electronics, Samsung's diverse business portfolio extended to other sectors, including life insurance, securities, and trading. This multifaceted approach showcased Samsung's presence in both technology-driven industries and broader financial markets, illustrating the company's prominence across various sectors of the global economy.

The rise of Samsung Electronics

Samsung Electronics, founded in 1969, has positioned itself as one of South Korea's premier manufacturers and has achieved global recognition. In the 1980s, Samsung established a new corporate identity centered around securing proprietary technology, improving research and development efforts, and expanding its semiconductor business. The significant moment came in 1988 when Samsung Electronics merged with Samsung Semiconductor & Telecommunications, marking its entry onto the global stage. By 2007, achieving $100 billion in sales, Samsung had positioned itself among the top three consumer electronics firms worldwide. A watershed moment occurred in 2010 when it gained the title of the world's largest electronics company by sales, firmly establishing its dominance in the global electronics industry. Continuing its commitment to innovation, Samsung has maintained its leadership in memory semiconductors, TVs, and smartphones. In 2019, the company introduced a new Corporate Social Responsibility (CSR) vision *"Together for Tomorrow! Enabling People."* This strategic vision outlines Samsung's dedication as a global corporate citizen, articulating its commitment to fulfilling social responsibilities and establishing the guiding ethos for the company's trajectory into the next century.

Leadership Responsibility and New Environmental Strategy

The mission of Samsung Electronics in 2023 is reflected in the following words: *"Our promise to perform responsibly as a leading, global company."*

Samsung Electronics leadership has a high level of awareness about environment, and responsibility that the company has in its daily operations. Jong-Hee Han, Vice Chairman and CEO of Samsung Electronics said: *"The climate*

[6]Ensar Mekic, Sultan Qaboos University. Based on the bibliography used.

crisis is one of the greatest challenges of our time. The consequences of inaction are unimaginable and require the contribution of every one of us, including businesses and governments."

He also added that Samsung is responding to the threats of climate change with a comprehensive plan that includes reducing emissions, new sustainability practices, and the development of innovative technologies and products that are better for our planet.

Samsung Electronics Announced a New Environmental Strategy in Korea on September 15, 2022, which is centered around three broad objectives:

- To achieve net zero emissions by 2050; Joins the RE100 initiative.
- Energy- and resource-efficient products to empower consumers for sustainable living.
- To invest in and develop innovative technologies for a better planet; Apply carbon capture and clean air technology.

The adoption of this new strategy represents more than just a formal document for the company; it reflects a profound orientation. In order to effectively implement this strategy, the company has developed implementation roadmaps for each environmental goal, encompassing targets such as achieving net zero and advancing the circular economy. To ensure the execution of these plans, progress tracking, and implementation are entrusted to the Sustainability Council, chaired by the CEO, and the Sustainability Committee, comprising external directors. This strategic approach underscores the company's commitment to realizing its environmental objectives through a structured and accountable framework.

Samsung Electronics Initiatives for 3Ps

Samsung Electronics, as a corporate citizen, is dedicated to the principles of the 3Ps:

- **Planet:** The company is committed to embedding environmental sustainability in all its products, aspiring to take a leadership role in fostering a sustainable future.
- **People:** Samsung places a paramount emphasis on creating a better life for all. Recognizing the value of people, the company actively pursues co-prosperity with its partners, fostering collaborative relationships that contribute to the well-being of individuals and communities.
- **Principle:** Samsung Electronics is dedicated to cultivating a transparent and healthy corporate culture. This commitment is grounded in the principles of compliance and ethics, reinforcing the company's pledge to uphold high standards of integrity in its operations.

Planet Oriented Initiatives of Samsung Electronics.

Numerous initiatives undertaken by Samsung Electronics toward environmental sustainability are depicted in Table 9.2.

People-Oriented Initiatives of Samsung Electronics

The company achieves responsibility toward people using a Human Rights Framework presented in Table 9.3:

Principle-Oriented Initiatives of Samsung Electronics

At Samsung Electronics, adherence to laws and ethical standards stands as a paramount business principle. The company is committed to conducting its business in strict compliance with both legal requirements and ethical standards. This commitment is driven by the desire to fulfill social responsibilities and foster a corporate culture that is characterized by fairness and transparency. This is reflected in the words of Jungho Park, Chief Compliance Officer (CCO) who said: "We foster a culture deeply rooted in compliance through continuous review and active practice."

Compliance and Ethics Program is realized through three important aspects: Prevention, Monitoring, and Response.

Galaxy S23 Ultra as a Great Example of Sustainability

At Samsung Electronics, the environmental impacts of their products are comprehensively evaluated throughout the entire life cycle. This assessment includes the sourcing, production, distribution, product use, and recycling phases. If we pay attention to the product Galaxy S23 Ultra, the one can observe the following details:

- Twenty percent of the plastic parts by weight are derived from repurposed plastic sourced from discarded fishing nets and water barrels.
- Twenty-eight percent of the aluminum used by weight in the Galaxy S23 Ultra is recovered scrap obtained during the manufacturing processes.
- The device is the first to incorporate Corning Gorilla Glass Victus 2, which contains an average of 22% pre-consumer recycled glass for the front screen and back cover.
- All business sites involved in the production of the Galaxy S23 Ultra have achieved ISO14001 (environmental management system) and ISO50001 (energy management system) certifications.
- Galaxy S23 Ultra contains only 1.6% of plastic in terms of the total weight of packaging.
- Compared to the Galaxy S7, the use of plastic in the packaging for the Galaxy S23 Ultra has decreased by 96.8%.
- The packaging box itself is made from 100% recycled paper.

Discussion Questions:

1. What is the role of Samsung Electronics Leadership in Sustainability of the company?
2. Does Leadership of Samsung Electronics implement TBL? How?

3. Do Samsung 3Ps correspond to 3Ps from TBL approach? How?
4. How Sustainability Council support leadership in implementing TBL?

Bibliography

Family Business magazine. (2004). *The World's 250 Largest Family Businesses*. Retrieved December, 2023 from www.griequity.com: https://www.griequity.com/resources/industryandissues/familybusiness/top-global.html

Samsung. (2023). *Samsung Electronics Sustainability Report 2023*. Samsung, Samsung Electronics. Samsung.

Samsung. (2023). *www.samsung.com*. Retrieved December, 2023 from Company Info: https://www.samsung.com/ae/about-us/company-info/

Samsung. (2023). *Our Journey Towards A Sustainable Future*. Retrieved December, 2023 from www.samsung.com: https://www.samsung.com/global/sustainability/

Samsung. (2023). *Compliance & Ethics*. Retrieved December, 2023 from www.samsung.com: https://www.samsung.com/global/sustainability/principle/compliance-ethics/

Samsung. (2023). *Human Rights*. Retrieved December, 2023 from www.samsung.com: https://www.samsung.com/global/sustainability/people/human-rights/

Samsung. (2023). Product Environmental Report: Galaxy S23 Ultra. Samsung, Samsung Electronics. Samsung.

Samsung. (2022, September 15). *Samsung Electronics Announces New Environmental Strategy*. Retrieved December, 2023 from https://news.samsung.com: https://news.samsung.com/global/samsung-electronics-announces-new-environmental-strategy#

Table 9.2 Planet-oriented initiatives of Samsung Electronics

Initiative	Implementation
"0" Carbon Emissions by 2050	Samsung Electronics is committed to achieving a substantial reduction in carbon dioxide equivalent (CO_2e) emissions, amounting to approximately 17 million tons, as indicated by the 2021 figures.
Ultra-Low Power Products	In line with its dedication to a more sustainable planet, Samsung Electronics is actively ensuring that its products prioritize energy efficiency, resulting in reduced electricity consumption.
Resource Circularity	Samsung Electronics upholds its commitment to sustainability across the entire product lifecycle, embracing responsible practices from the sourcing of raw materials to the eventual stages of product disposal and recycling.

(continued)

Table 9.2 (continued)

Initiative	Implementation
Ultra-Low Power Semiconductors and Energy-Efficient Electronics Products	Samsung Electronics is strategically incorporating new low-power technologies to curtail energy consumption in everyday consumer electronics. This initiative encompasses the development of cutting-edge ultra-low power memory chips.
Maximizing Resource Circularity Across the Entire Product Lifecycle	Samsung Electronics is intensifying its commitment to enhance the resource circularity of electronics throughout the entire lifecycle of a product. The aim is to optimize resource utilization, ensuring that every material is employed with minimal environmental impact.
Water Conservation and Pollutant Treatment Measures	With the anticipated expansion of domestic semiconductor manufacturing capacity, the company foresees a doubling of daily water withdrawal needs from its semiconductor operations in Korea by 2030. However, in a commitment to sustainable practices, Samsung Electronics is pledging to maximize water reuse.
Invest in and Develop Innovative Technologies for a Sustainable Future	Samsung Electronics is harnessing its cutting-edge technology to tackle global climate challenges. The company is placing a specific emphasis on the development of carbon capture and utilization technologies.
Carbon Capture Research Institute	Samsung Electronics established the Carbon Capture Research Institute within the Samsung Advanced Institute of Technology (SAIT) in September 2021.
Clean air technologies	This initiative involves the creation of new filtration systems aimed at reducing particulate matter.

Table 9.3 Human rights framework of Samsung Electronics

Item	Implementation
Policy	Samsung Electronics is committed to treating all stakeholders, such as customers, executives, employees, partners, and local communities, with fairness and respect for their rights. The company believes that upholding human rights is not just a moral obligation but is also essential for enhancing the overall value of its business. This commitment reflects Samsung's dedication to positively contributing to the well-being of everyone associated with the company.
Due Diligence of Human Rights	Samsung Electronics established a sustainable due diligence system to accommodate social, geopolitical, and environmental changes, such as the outbreak of the COVID-19 pandemic.
Remedial Action	The company operates various grievance channels for internal and external stakeholders tailored to the conditions of individual business sites.
Transparent Disclosure	Samsung Electronics provides detailed reporting on how it implements human rights commitments across business activities in our Sustainability Report based on the Global Reporting Initiative standard, as well as other standards.
Stakeholder Communication	The company remains committed to ensuring open and active communication with stakeholders based on mutual trust.
Governance	Samsung Electronics operates a Global Labor Issue Committee to oversee labor and human rights impact/risk management. Executives and working-level employees in the eight divisions of the Global Labor Issue Committee participate and consult on labor and human rights issues twice a month. Any critical matters are escalated to the Sustainability Council, which consists of key executives, and then reported to top management and approved by the Board of Directors.

As a family business grows, it encounters more complexity and challenges. Yet, with skilled leaders, all these challenges can be effectively handled, provided the family firm leaders apply the framework mentioned earlier, Fig. 9.3.

Questions for Discussion

- Discuss briefly TBL in family business.
- Discuss the nature of family business leadership within the TBL framework.
- Explain the implications of family leadership within the TBL framework.
- How the family firm will succeed to implement TBL while addressing internal issues?
- Why family firms can outperform SMEs in their TBL endeavor? Provide valid examples.

Additional Readings

- Ramadani, V., Memili, E., Chang, E. P., & Palalić, R. (2020). *Entrepreneurial family businesses: Innovation, governance and succession* (1 ed.). Springer.
- Ramadani, V., Bexheti, A., Rexhepi, G., Ratten, V. & Ibraimi, S. (2017). Succession issues in Albanian family businesses: Exploratory research. *Journal of Balkan and Near Eastern Studies, 19*(3), 294–312.
- Dana, L-P. & Ramadani, V. (2015). *Family businesses in transition economies*. Cham: Springer.
- Miller, D., & Elkington, J. (2015). *Can the world be wrong: Where global public opinion says we're headed*. Routledge.
- Elkington, J. (1997). *Cannibals with forks: The triple bottom line of 21st century business*. Oxford OX2 OJX.

Suggested Activities

1. Find two firms, one an SME and the other one family firm. Identify the main difference. Bring the triple-bottom-line approach intensity in both types of firms and identify which one performs better. Justify reasons in both cases (doing less; doing more).
2. Conduct an interview with a family business owner and identify the strategy for the firm's longevity.
3. Select two or three family firms in your region and try to find out the succession plan that will lead to long-term sustainability.
4. Identify two family firms and compare them in terms of how they address the new global agenda on the triple-bottom-line approach.

Keywords and Concepts

- Family businesses
- Family businesses leadership
- TBL in family business
- STEP in family businesses
- Transgenerational Entrepreneurship
- Entrepreneurial orientation in family business
- Familiness
- Family business ambidexterity

References

Websites

FORBES.: https://www.forbes.com/sites/forbescoachescouncil/2020/03/18/strategies-for-family-business-success/?sh=3af6547e71c7. Accessed on 20.08.2023.

http://millsproperty.com/about-3/. Accessed on 27.02.2020.

https://fambus.org/mills-property-management/ Accessed on 27.02.2020.

The Family Business Consulting Group. Available at: https://www.thefbcg.com/Secrecy%2D%2DA-Family-Business-Vice/. Accessed on 21.02.2020.

Ahmad, N., Smajić, H., Palalic, R., & Al-Riyami, S. (2022). Sustainability management research in the Arab world: A citation-based systematic literature review. *Management and Sustainability: An Arab Review, 1*(1), 14–37. https://doi.org/10.1108/msar-01-2022-0004

Basco, R., Calabrò, A., & Campopiano, G. (2019). Transgenerational entrepreneurship around the world: Implications for family business research and practice. *Journal of Family Business Strategy, 10*(4). https://doi.org/10.1016/j.jfbs.2018.03.004

Berrone, P., Cruz, C., & Gomez-Mejia, L. R. (2012). Socioemotional wealth in family firms: Theoretical dimensions, assessment approaches, and agenda for future research. *Family Business Review, 25*(3), 258–279. https://doi.org/10.1177/0894486511435355

Calabrò, A., Torchia, M., Kallmuenzer, A., Yezza, H., & Feng, C. (2022). Transgenerational entrepreneurship in family firms: a configurational approach. *Review of Managerial Science, 17*(7), 2535–2554. https://doi.org/10.1007/s11846-022-00592-y

Debicki, B. J., Kellermanns, F. W., Chrisman, J. J., Pearson, A. W., & Spencer, B. A. (2016). Development of a socioemotional wealth importance (SEWi) scale for family firm research. *Journal of Family Business Strategy, 7*(1), 47–57. https://doi.org/10.1016/j.jfbs.2016.01.002

Etemad, H., Wright, R. W., & Dana, L. P. (2001). Symbiotic international business networks: collaboration between small and large firms. *Thunderbird International Business Review, 43*(4), 481–499.

Gómez-Mejía, L. R., Haynes, K. T., & Núñez-Nickel, M. (2007). Socioemotional wealth and business risk in family-controlled firms: Evidence from Spanish olive oil mills. *Administrative Science Quarterly, 52*(1), 106–137.

Gómez-Mejía, L. R., Cruz, C., Berrone, P., & DeCastro, J. (2012). The bind that ties: Socioemotional wealth preservation in family firms. *Academy of Management Annals, 5*(1), 1–79.

Habbershon, T. G., Nordqvist, M., & Zellweger, T. M. (2010). Transgenerational entrepreneurship. In M. Nordqvist & T. M. Zellweger (Eds.), *Transgenerational entrepreneurship. Exploring growth and performance in family firms across generations* (pp. 1–38). Edward Elgar Publishing.

Kalm, M., & Gomez-Mejia, L. R. (2016). Socioemotional wealth preservation in family firms. *Revista de Administração, 51*(4), 409–411. https://doi.org/10.1016/j.rausp.2016.08.002

Longenecker, J. G., Petty, J. W., Palich, L. E., & Hoy, F. (2017). *Small business management* (18th ed.). Cengage Learning.

Merck Group. (2023). Accessed December 20, 2023, from https://www.merckgroup.com/en/company.html

Palalić, R., & Smajić, H. (2022). Socioemotional wealth (SEW) as the driver of business performance in family businesses in Bosnia and Herzegovina: The mediating role of transformational leadership. *Journal of Family Business Management, 12*(4), 1043–1064. https://doi.org/10.1108/jfbm-07-2021-0067

Ramadani, R. P., Dana, L.-P., Krueger, N., & Caputo, A. (2020a). *Organizational mindset of entrepreneurship*. Springer. https://doi.org/10.1007/978-3-030-36951-4

Ramadani, V., Memili, E., Chang, E. P., & Palalić, R. (2020b). *Entrepreneurial family businesses: Innovation, governance and succession* (1st ed.). Springer.

Ramadani, V., Memili, E., Palalić, R., & Chang, E. P. C. (2020c). Entrepreneurial family businesses. *Springer*. https://doi.org/10.1007/978-3-030-47778-3

Rothaermel, F. (2017). *Strategic management* (3rd ed.). Mc Graw Hill Education.

Zellweger, T. M., & Astrachan, J. H. (2008). Building a family firm image: How family firms capitalize on their family ties. *Family Business Review, XXI*(3), 347–363.

MIX
Papier aus verantwortungsvollen Quellen
Paper from responsible sources
FSC® C105338

If you have any concerns about our products,
you can contact us on
ProductSafety@springernature.com

In case Publisher is established outside the EU,
the EU authorized representative is:
Springer Nature Customer Service Center GmbH
Europaplatz 3, 69115 Heidelberg, Germany

Printed by Libri Plureos GmbH
in Hamburg, Germany